Unireality

Stanley Heidrich

In memory of Albert, Carl, Julian, and Stanislaw

and to the dream of a peaceful future

this book is dedicated.

Published by Stanley Heidrich

ISBN 978-0-6151-4814-4

Contents

Contents

An Introduction to Unireality

Reality is a strange and wonderful thing. It surrounds us, while also being inside us, being our thoughts and deeds, our substance and our spirit. Reality is the road that we trudge down and the classroom where we learn, it is both ego and sense of self, it is history, science, and literature, and it is religion. Reality is wherever a person looks for it and also everywhere a person doesn't look for it. Reality has a thousand faces and reveals a new one whenever the curious person seeks novelty, and yet it is always the same.

Reality can be imagined as a strange creature or a complex machine, as a glorious tableau or a construct created by extension of a four-dimensional fractal algorithm. But however one imagines reality, it should be understood that it is a thing. Reality is a thing-in-itself, an entity, a holism. What it is *not* is a collection of things, a scattershot agglomeration of unrelated bits. Reality is complete, and if one bends one's mind to the task it is comprehensible. Whether living, machine-like, picturesque, or mathematical, it behaves predictably, even when it is being unpredictable. Reality is something that can be studied.

Over the last few thousand years, a great deal has been spoken and written about what happens in reality. The things and events of reality have been picked apart, sifted through, analyzed, sorted, filtered, and distilled. Patterns have been derived and claims made. Sometimes the patterns are illusions and dangerous, sometimes they are mostly accurate and useful. However, even with the actual laws of physics, which describe how events happen at the simplest levels, the patterns that have been derived tell us almost nothing about what reality is. They only tell us how to control it, in the simplest and crudest fashion.

Then what is reality? The most accurate answer is that reality is a way of looking at Nothingness so that Somethingness seems to be there. It may be more economical to say that reality is a way for

Nothingness to view itself. We observers, of course, are part of reality. That means that we are eyes through which Nothingness sees itself.

To embrace this definition, one must get used to the idea that our universe sums up to zero—that it is equivalent to Nothingness. That idea is the core concept of this book. It isn't a brand-new notion; for several decades physicists have entertained the possibility that our universe may be some kind of vacuum fluctuation. Nor is the idea far-fetched: the laws of quantum physics allow a universe of zero total mass/energy to exist indefinitely.

There are very good reasons to believe that our universe is, in fact, zero-sum. There are also very good reasons for believing that, if our universe is equivalent to Nothingness, it is emerging from the Nothingness continuously, and according to a single transformational principle. In other words, there is a way, a technique, a single principle by which Somethingness is being made out of Nothingness. And, there is very good reason to believe that that single principle governs everything happening in our universe.

* * *

To my knowledge, there has never before been a book written about zero-sum reality. Without a precedent, there is no standard template to follow. Reality isn't a subject like the Russian Revolution, or global warming, or the Mayan calendar. It doesn't fit into any particular genre or discipline. It's not exactly physics, and not exactly philosophy. It draws deeply on history, economics, cognitive science, and religion, but is too expansive for any one alone. Metaphysics or epistemology comes closest, but those are more like subheadings, aren't they? No subject is broader than reality.

When one sets out to write such a book, there's another difficulty. The language, ideas, and the creative process itself are all part of the subject. Writing about reality is a bit like trying to capture the image of a big mirror with smaller mirrors.

Of course, I didn't set out to write a book about reality. This all began when I was twenty years old and trying to devise a theory of how human cognition functioned. Or perhaps it began on my fourteenth birthday, when an aunt gave me Bertrand Russell's book on the theory of relativity. The subject fascinated me, as did quantum physics a decade later. But there were so many subjects to investigate, and before long I abandoned my theorizing on human cognition, in favor of a new inquiry into the changes in human consciousness through history. All the while I read and read, pausing once in a while to synthesize in essay form the ideas that I had jotted down while reading. But these essays were about political science, about economics, about psycholinguistics and ethics. They were not about reality itself.

One late night as I was immersed in a volume of Durant's history, an elusive recognition teased at my mind. There was the faintest hint of similarity in the way events were being described by him to something else I had read. Not that I could put a finger on it. But over the next months, that elusive sense of some pattern hiding in the tapestries of reality recurred more and more frequently. One day when I was reading about quantum physics, it dawned on me that the quantum phenomenon of correlation, or phase entanglement, was essentially the same as the interpersonal connection we call love. But that was crazy. Why would quantum physics have any phenomenon in common with something as non-physical as love? Unless...

* * *

I went back and re-read many of the books I'd read before, and mined them for more connections, more similarities. With little success. But as I searched for such patterns, something did begin to nag at me. It was a problem with symbols. At the time I was trying to reconcile what I imagined to be actual reality with what I was calling symbolic reality. I had become suspicious that human problems were caused by flaws in the way physical reality was represented with abstract symbols. Here I was influenced by the Sapir-Whorf reflections on linguistic biases. Perhaps, I thought, the blame so often put on human nature wasn't justified. Perhaps our symbolic representations were being clumsily used, and causing us our grievances. But whenever I tried to specify the definition of the term *symbol*, it slipped from my grasp. And finally the reason dawned on me—*everything* was symbolic. There was no dividing line between actual and symbolic reality. Everywhere one looked, all one saw was symbols. It wasn't just words and musical notation and pictographs and mathematics terms and traffic signs and cultural icons. It was atoms and molecules, chromosomes and genes, rivers and valleys, hurricanes and tornadoes, living beings. Symbols were packages of information, summaries of the past. They were like memories of what had happened. Because—this was a realization of great clarity—the process of reality was one of building symbols through accumulated observations.

Unfortunately, the relief and joy that came with this epiphany were short-lived. There was still so much I didn't understand about how physical reality worked, and I couldn't very well show that physical and abstract reality worked similarly until I did. Plowing back into physics, I scoured the field for a starting point from which to work. However, physics is little more than a collection of *ad hoc* concepts invented to explain various empirical data. Physicists don't really know what anything is. They don't know what energy is, what an observation is, or what an atom is; they only know how the phenomena work. It's astonishing. Physicists can't even say why our universe has dimensions, and they can't begin to explain what time is. I wandered in their Lewis Carroll wilderness for a long time, trying to make sense of it.

Relativity theory and quantum theory have undeserved reputations for being hard to understand. The real bogeymen in physics are the most basic concepts. What is a vibration? What is a wave? What is a particle? What is a dimension? Not what do these things look or act like, but what *are* they? Where do such things come from, and how do they fit together to make our reality? There aren't satisfactory answers in the physics books.

I was sitting in my old blue overstuffed chair one evening, sipping black tea, trying to figure these things out for the thousandth time, when the visual images I sought finally fell into place. Wasn't a wave just a vibration with another spatial dimension to move in? And if it had another spatial dimension to move in, a wave became a forward spiral, a helix. And, if that vibration had its one spatial dimension collapsed, it became a blinking dot. I scribbled sketches in the yellow legal pad laid on the arm of my chair.

Didn't that solve the wave-particle duality problem? Of course it did! The phenomenon we think of as light could express itself in different guises, depending on the number of dimensions that we measured it in. A measurement is an *observation* in physics, and making an observation collapses a wave function. And, when a wave function collapses, a correlation is broken. Couldn't we say then that observations break correlations by collapsing dimensions?

But what was a dimension? There were four of them, that was clear enough. And by this time it seemed very apparent to me that each dimension was a zero-sum construct, made by separating Nothingness into a positive and negative value. I was even certain that we had four fundamental forces because we had four dimensions. No matter how hard I tried, though, I couldn't visualize how those four binary dimensions combined to create our universe.

I was in LAX airport, waiting between flights, gazing at taxiing planes, when the image suddenly materialized in the air above the tarmac—more vivid and bold than any abstract image I had ever conjured up. I stared at it out there in all its glorious wonder, a convoluted pretzel-like construct of four manifolds flowing in and around each other to make a three-dimensional representation. It was the most marvelous thing, an elegant and complete answer to the dimensional problem. In that singular image I saw the essential structural process of the universe. I saw the key to unlocking the secrets of reality that had resisted me for so long. What I saw was the interface between Somethingness and Nothingness.

* * *

This is a book about reality. It is a story about the wonderful and mysterious place that we find ourselves birthed into, and in which we wander for a handful of decades before leaving for parts unknown. It's a place about which trillions of words have been spoken and written, as men and women down through the millennia have sought to understand it well enough to meet their needs. Some small percentage of those trillions of words are composed in the form of questions, asked of no one in particular, at moments of undistracted self-awareness. Why am I here? What is the purpose of life? Does any-thing matter?

The answers to these questions won't be found in this book. There aren't stock answers to them, because such questions make no sense outside a stated context. A mother with a hungry baby knows why she is here; a man on a mission knows the purpose of his life; an artist concentrating on his or her latest work knows exactly what matters. But in the broadest possible context, in the light of eternity—which is how most people who pose these questions mean them—the questions themselves are meaningless. A person is here because the person can be here. The purpose of life is life. Everything matters and nothing matters. The questions can have meaningful answers inside reality where they belong, but the Nothingness shrugs them off as meaning-less.

Another common question of the type is this one: who or what caused the world to be? Who or what created it? The implication of this teleological question is that the world can't exist without having been made; after all, everything that is or happens has a cause. Like those other questions, this one only makes sense inside reality, and not outside it. Causality is a feature of our universe, and is fully ex-plicable. But causality isn't a feature of the Nothingness. Nothing-ness doesn't have any features. It has no dimensions, and it has no contents—although it can be anything at all, as long as the anything is equal to nothing. So, a first cause is unnecessary. In fact, if our uni-verse is zero-sum, it cannot have had a cause. It exists because it can exist, not because it was made to exist.

That philosophical questions of this kind lose their meaning at the interface of Somethingness and Nothingness doesn't mean that any-thing goes, that anything is permissible, that morality is phony. Mo-rality does tend to be culturally relative, but reality most assuredly is imbued with a referential morality that is implicit in its dynamic structure. The morality isn't handed down by an authority on high. It is in the way things work. The universe is governed by the condition that it stays zero-sum at all times, and this condition enforces on all events an implacable and inescapable control. In a universe without such a condition—in the kind of universe that we are *told* we live in—there is no ultimate control; any morality must have been arbi-trarily invented, either by a god or by us. In a universe without such a condition, a chain of events can travel in any direction, ungoverned by anything but elementary laws of physics, and reach a consequence affected by the originating event only by the links of the chain, which have been chosen arbitrarily or probabilistically. That cannot happen in our universe. No matter how arbitrary or probabilistic the links in the chain of events may seem, there is that primary condition invisibly helping to select them. In our universe no events are entirely random. Even though this makes us more responsible than we would be in a universe where randomness is possible, it also assures us that our universe makes sense, which is what we mean when we say it has an implicit morality.

Because this is a book about reality, and reality makes sense, values are assigned to some kinds of behavior and courses of action. This isn't done with the intention of moralizing, at least not in the usual sense of the word, but rather to identify the actions in terms of their negative consequences. My intention at all times is to describe reality as accurately and comprehensively as possible, in a nuts and bolts manner. Philosophical and moral opinions belong in a different kind of book.

* * *

This book can be summarized in a single sentence: *Our universe represents Nothingness through a process that expresses itself in all things and change.* That is not a small statement. Presupposing that it can be supported logically and/or empirically, it invites an array of interpolations and extrapolations, with implications leading in all directions. However, the further one travels in any single direction, the deeper one goes into forest, and the harder it is to see anything but trees. This isn't a trivial problem for the writer or the reader.

Furthermore, although reality can be easily described at its most fundamental levels, describing it at higher levels is made a formidable task by the complexity of emergent structures. Although all these structures are designed and governed by the same dynamics, they do have different aspects to them, and the aspect that serves best for one purpose may not work as well for another.

Since these problems can't be avoided they might as well be embraced. That is what I have done, including selected works that trace the investigatory path I followed. They don't have to be read in the order I have presented them. In fact, the advised approach is for the reader to find an accessible and interesting part and start there.

At the end I have included a commentary on how the world today looks, from the point-of-view of a unirealist. The focus is naturally on the United States, with an eye to what must be done if we want to restore health to its economy and body politic. I did strive here to minimize abstract theorizing—with mixed success.

"Theory of Unireality," which comprises the long Part Three, is a comprehensive and wide-ranging series of conversations about the underlying nature of reality. These conversations are from twenty years ago, and there is some that I would change about what I said in them. Nevertheless, they marked the peak of my creativity, and since then my efforts have gone toward developing and clarifying the ideas they initiated and expressed.

Part Two is comprised of four academic papers and two more casual ones. These were all written during the same recent period of time. Two are about physics and two about cognition. Another is about what I call metasystems. The last is about intimate relationships. The first paper on physics explains why our universe exhibits relativistic distortions, the second why light can appear to be either particle or wave. The two papers on cognition explain and discuss awareness, consciousness, thinking, and hemispheric specialization.

Part One is titled "Understanding Reality." Like the earlier work "Theory of Unireality," it develops the paradigm in which reality is described as a thing-in-itself. However, while "Theory of Unireality" primarily discusses the symbolic structure of reality, this more recent work considers reality in terms of the systems made of those symbolic structures, with emphasis on how such systems exchange energy. The two approaches are compatible, but "Understanding Reality" offers theoretical precision, as was my intent in writing it.

For the most part, the ideas expressed in these pages are my own. I have drawn extensively from general scientific knowledge. However, wherever I borrowed an idea or concept from a particular person, I have given that person credit in text or in a footnote.

* * *

There is a last question that has come up, and this is a good place to address it. That question is, how complete a conceptualization of reality is this? Can we chuck all other philosophical and metaphysical speculations out the window? And, since I have proposed that all reality conforms to a single set of transformations, are other abstract laws and formal approaches now irrelevant? My friend Tom poses this query, and it is a good one that I had not myself considered, but I'll try to respond here at the end—or the beginning, if you will—as efficiently as I can.

Tom, I am not enamored of philosophy and metaphysics, at least not as practiced. Marx insisted that all the pointless ivory tower philosophizing had to end, and more attention be paid to what was going on in the material world economically and politically. That isn't my objection, but he did have a point. Far too much effort goes into assigning names and creating categories and far too little into understanding the actual processes of reality. Philosophers tend to *thing-ify* reality to the point of mind-numbing specificity. The *things* in reality are important only insofar as they are understood as expressions of *process*. Marx and Engels realized this. The essence of dialectic materialism is change itself—development, evolution, motion, struggling—and its role in human society. Marx and Engels tried to formulate the natural laws that governed the world. Their approach to reality was a dynamic one, which helps explain its power to inspire popular movements. In that respect, they were like Freud, who earned his influence by emphasizing the dynamism of the human psyche—the active processes that made it what it was. Other psychologists, meanwhile, fell into the same trap as other philosophers, using up their energies naming things and categories and attitudes and then trying to connect them together relationally in a static framework.

Please realize that I am not especially enamored of Marx and Freud, either. Freud's penchant for inventing things out of whole cloth was egregious; what else are the ego, id, and superego? With Marx, at least his things had substance; capital, machines, and industry are not abstractions. But the dynamism of his approach was borrowed from Hegel and Engels, and lost its idealism as well as its usefulness in the borrowing. The inquiry into nature's general principles of change got lost along the way. As the disastrous era of institutionalized communism began, the idea of process in reality was perverted into the blasé lie that the world was evolving into a worker paradise.

The world is dynamic and things do change in predictable ways, but apart from the now-tarnished dialectics of Engels, there has been surprisingly little effort made to understand what the ways are and why they are so deeply embedded in reality. There has been a tremendous intellectual failure in this regard. There are plenty of assertions made today about how X predictably leads to Y, but they are usually made by political pundits, and are more bluster and bluff than serious attempts to understand reality. All the brainpower and dollars invested in thinking about reality, and the best that we can offer future generations, insofar as describing how things change in reality, are a few sage but simple rules of thumb: Murphy's Law, Gresham's Law, the Peter Principle, the Tipping Point, and so on.

So, yes, most other philosophical and metaphysical theorizing should be chucked out the window. Most philosophical thought is desiccated and barren; it has nothing to do with anything at all of importance, and it certainly doesn't help us understand our world any better. For centuries the philosophers have conjectured about causality, while here I have actually explained it. For centuries the philosophers have struggled to bridge the gap between materiality and immateriality, while here I have made that gap disappear. For centuries the philosophers have attempted to categorize and describe the various types of reality, while here I have shown that there is only one type. If my description of reality proves correct, previous schools of philosophy are either irrelevant or need dramatic revision. The library table is swept clean, and philosophers can begin their work anew.

Tom, what I have done is built an anchored framework. I'm not even sure I can explain any longer why I did it, or even how. The effort may have been driven by a quirk in my cognition, which is that I have to be able to visualize a concept to use it. Most people get through life perfectly well without being able to visualize the concepts of energy, consciousness, dimension, symbol, and all the others I have explained in these pages. I can't. My brain wants to be able to "see" things. All the serious thinking I do is done with images. What is marvelous about an image is that if it can be successfully realized in the brain, it will also be consistent with reality. The brain actually constructs a three- or four-dimensional neuronal model when it creates an image of something that in reality has three or four dimensions. Quite amazing, really. Anyway, in order for me to understand something like quantum physics, I had to be able to visualize those physics terms, and I couldn't visualize them until I actually understood them and my brain could model them. Similarly with history, economics, psychology, and so on. In order to learn them, I had to be able to visualize the concepts in each field, and to do that I had to make accurate models of them. I've done mental work like this for so long that I can visualize almost anything, even four-dimensionally, if it is something existing in four dimensions, like a system. And then this funny thing happened on the way to the forum, as I explained earlier: I realized that abstract visuals in one discipline were sometimes similar to, or resonated with, abstract visuals from another discipline. I worked long into the nights developing and integrating those visuals, and voila, here I am now with, as I said, an anchored framework.

Which is to say, there's still a lot of work to be done, but at least now we have a place to start. I have located the interface between Somethingness and Nothingness. From that epistemological foundation I extracted the fundamental processes of reality, while explaining what the most common terms in physics meant. But I suspect that as one ascends to higher levels of complexity in our universe, emergent dynamics and properties appear, and these I haven't even begun to explore or address. So I don't think the theory is complete. I don't think our understanding will ever be complete. Reality is a strange and wonderful thing, and I doubt it will ever bore us. It's never bored me.

You were a friend when I needed one, Tom. Thanks for everything.

* * * * *

Part I

Understanding Reality

1. A Single Reality

If we want to understand reality, we should begin by believing that reality is understandable. But notice the wording there. Casually read, it might seem to suggest that everything *in* reality can be understood. Not so. Rather, the statement says that reality *itself* can be understood—that there is a single reality, and it is understandable as a thing-in-itself.

Let's take just a moment here to consider how different a way of looking at things that is. Our species only extends about a hundred thousand years into the past. We know very little about the thinking processes of hominids that preceded and presumably evolved into us—Homo sapiens. In fact, we know little about how Homo sapiens thought, prior to the invention of phonetic writing, which was about three thousand years ago. We know that societies before then maintained village settlements with agricultural systems and even small cities. They had rituals and religions, they ceremonially buried their dead, they created art, they warred and traded. However, we don't know what they thought about or *how* they thought. It isn't really until the heyday of the Ancient Greeks that substantial evidence indicates thinking of the kind we engage in. And, we know that they—Plato and Socrates and Archimedes and Democritus and Pythagoras—devoted a great deal of time to trying to understand reality. What seems really curious today is that these Greek philosophers believed that reality could be understood in a single way. They thought of reality as a single unbroken whole, and from many different angles they tried to discern its true nature. These different approaches became what we call our academic disciplines: geometry and biology and mathematics and history and physics and chemistry and psychology. But over these past two thousand years we have abandoned that central concept in Greek philosophy—that reality is a unity and can be studied as a thing-in-itself.

No one anymore believes reality is that simple. Oh, it is true that physicists hope to discover a Theory of Everything, but what they mean by this is a unification of the fundamental equations of physics. They don't expect this anticipated TOE to suggest anything about human psychology or history. But if reality is comprehensible it is also consistent, in all areas and across all disciplines. Otherwise it is not comprehensible *as* reality. One can have it one way or the other but not both: either reality is a catchall term for all sorts of things, many of which we understand, or reality is the way things are, in which event we can understand it as a consistent pattern of structure and behavior. And if it is used in this latter sense, it applies to all things—whether they are concrete or abstract. Usually people suppose that the term reality refers to physical reality, so if economic reality is being discussed, or spiritual reality, or psychological reality, then the appropriate prefix is specified. That we require such distinctions implies that we believe physical reality and non-physical reality to be different, as if the non-physical realm emerges from the physical realm but has its own operative principles. To be sure, we do employ terms from physics—inertia, power, leverage, light, momentum, etc.—to describe events and processes in non-physical milieus, but we only use the terms metaphorically. But this should tell us that, at the very least, non-physical reality echoes physical reality. And if we think still more boldly it suggests that structures and behaviors in non-physical reality may be the same as in physical reality. Reality may be simple after all—consistent and comprehensible, the same everywhere.

2. Solid Footing

Because we don't believe in one reality, we're lost in a quagmire of unmeaning. We cling to our cherished beliefs, but we can't prove that any of them are true, so we make them axiomatic. Here we must be honest with ourselves. It doesn't really matter whether we are Christian or atheist, if we are progressive or conservative, if we believe in free markets, abortion rights, gun control, school vouchers, environmentalism, corporal discipline of children, the afterlife, or individualism or spiritual growth or anything else. The simple fact is that whatever one's belief may be, ultimately it is grounded in quicksand. There is no way to rationally establish that any such belief is undeniably valid. On some level we sense that, which is why we get defensive about our beliefs. And the fact that they can't be proven true explains why we use power—when rhetoric doesn't do the trick—to force the beliefs on others. The modern world is filled with power struggles—fistfights, acrimonious debates, conquests, and rebellions. That's how it has been ever since written history began and that's how it will continue, if we don't find solid footing for our systems of meaning, because there is no more effective way to prove right from wrong.

Is there anything we do know for sure? Back in the seventeenth century, Rene Descartes gave the question a lot of thought and decided the only thing we could be certain of was that we did in fact exist. But after beginning with this axiomatic truth he immediately erred. He decided that as we did exist, there had to be God who created us. How else could the world have been caused? From there he formulated Cartesian philosophy, with its pernicious mind-body split that still bedevils modern thought.

Yet if his first logical deduction was wrong—as is argued here—his initial assertion holds true. We *are* here, somehow and in some way, and it is the first thing we know. But what do we mean by *we*? Human beings? Sentient creatures? All living things? It's unclear. So let us re-state his axiom, and declare this: all that we know for sure is that there is a Somethingness. We call it a universe. We are in it, and part of it. Might the universe be an illusion? It might be, but there is still our awareness of the illusion, and in that awareness resides a Somethingness, whether it is merely an illusionary one or not.

If there is a Somethingness, there could also be not a Somethingness. That is, if we were not here at all, there would be Nothingness. Given the first concept, the second and opposite concept must also be allowed. And Descartes, in his analysis, realized then what we must realize now—that the relationship between these twin concepts is the source of all meaning, and that until it is made into a foundation all other statements about reality must perforce hang by skyhooks.

But Descartes, like so many philosophers before and since, thought the relationship causal. He jumped to the conclusion that Somethingness must have been caused to emerge from Nothingness, by a divine creator. Today we understand a different relationship is possible. The two concepts don't have to be connected causally, with one temporally prior to the other. They may simply be identities, existing simultaneously. In fact, such a relationship is more defensible than the causal one that Descartes took for granted. Divine creation runs contrary to basic laws of physics while equivalence does not.

If Somethingness and Nothingness are identities, then we are living in a zero-sum universe, and Somethingness is just another way of looking at Nothingness. This may seem a strange concept, but with it we can understand reality for the first time. We can see why reality works as it does, and thus find solid footing for our concepts, terms, reasonings, and beliefs.

3. The Language of Physics

To understand a thing, we must be able to express it representationally so that it fits into some paradigm. This can't be done without some kind of language, of course, because without a language one can't have representations or paradigms. The language doesn't have to be literal; there are non-literal languages such as mathematical, pictorial (like the diagrams of a football play), and musical notation. But all languages have this in common: they are made up of abstract terms that symbolize the actual elements of physical reality, and the relationships of those elements. The terms can be strung together, according to rules of grammar and syntax, to represent and describe discrete events in physical reality—as sentences of subjects and verbs, or equations, or diagrams. And these strings can then be assembled into a narrative about events in physical reality: maybe a story or mathematical proof or piece of music. The last step occurs when these miscellaneous narratives are melded into a paradigm. The paradigm is an attempt to generalize and unify, to make sense of the proliferation of narratives, to synthesize them into a comprehensive representation of multiple aspects of reality. So, when we say we understand something we mean that we can express it as a string of abstract terms in a narrative—a narrative that is consistent with the relevant paradigm, which is reasonably self-consistent.

Some paradigms in literal language are so extensive that they purport to encompass virtually all of reality. Marxism and Christianity are two examples. Most paradigms, however, are situation-specific. An auto mechanic has a paradigm for the workings of the internal combustion engine. If he is interested in politics, he also has a political paradigm, one that enfolds his knowledge and beliefs about political matters. He may very well have a religious paradigm, a family paradigm, and a military paradigm. And while these different paradigms are not mutually contradictory they are not unified, either. Our auto mechanic may respond to a given situation with different behaviors, depending on the paradigm with which he analyzes and evaluates it. Finally, such paradigms are plastic, and usually joined in a loose confederacy that closely parallels the individual (or cultural) sense of identity.

To discuss the nature of reality we must choose an appropriate paradigm. Since we are speaking of reality-as-a-whole (rather than some narrower aspect of it) our choices are limited to the three categories of religious, philosophical, and scientific paradigms. The last of these—specifically the paradigm of physics—works best for our purposes. In physics, there is a clear correspondence between the terms of its language and the phenomena in reality represented by those terms. Terms like *gravity* and *oscillation* and *atom*, for example, aren't subject to interpretation; they inflexibly refer to physical phenomena that can't be confused, or conflated, with any other phenomena. That isn't the case in religious or philosophical paradigms. Terms in religion, such as *faith* or *charity* (or the term *god* itself) are as fuzzy and subject to interpretation as terms like *intention* and *self* are in philosophy. The conceptual frameworks of religious and philosophical paradigms are poorly grounded, and it is impossible to confirm or refute or measure the accuracy of their statements about reality, faith and logic notwithstanding. Physics, on the other hand, is largely mechanistic. When it asserts a truth, that truth is subject to experimental confirmation.

Such being the case, one would expect that contemporary physics has an integrated and coherent paradigm of physical reality and that we can appropriate that paradigm wholesale for discussing reality. Unfortunately, it doesn't, and we can't. Physicists have failed miserably at developing a self-consistent paradigm. They have virtually ignored the all-important interface between Somethingness and Nothingness, and their current paradigm for explaining the nature of reality is filled with paradoxes.

4. The Zero-Sum Universe

Like Descartes, most people today—including many of our great thinkers—assume that our universe was somehow created. The idea that the Somethingness of our universe and the Nothingness that logically precedes it might be identities, and how such an equivalence would bear on the idea of creation, has not been significantly evaluated. Before exploring how this zero-sum reality of ours behaves we might briefly examine a few logical considerations.

To begin with, if a god perceives our universe from somewhere outside it, then our universe exists objectively. It is, after all, perception from outside that determines the objective reality of a thing. According to western religions our universe is objectively real in this way. However, a universe that exists objectively cannot be either zero-sum or self-enclosed; the only type of universe that can maintain a zero-sum and enclose itself is a universe that amounts to nothing and is not objectively real. This contradiction can't be reconciled. It's no use arguing that if God created this universe out of nothing it would be *ipso facto* zero-sum. Not only would God's creative venture introduce a bit of energy to the zero-sum equation and thereby upset it, but also any subsequent interaction God might have with the universe—even viewing it—would similarly introduce an energy imbalance. No, if God created our universe from the outside, then our universe exists objectively and cannot be zero-sum or self-enclosed. Technically, this wouldn't even be a universe, as the term literally means the totality of all that exists yet it wouldn't include God. Semantics aside, we nevertheless must accept that if God created the universe, it objectively exists but isn't zero-sum.

Why is this logical conclusion significant? Because, if our universe is like this, it can never make complete sense to us human beings who reside within it. Presumably God can understand it fully, but we can't. No matter how diligently we study the behaviors of reality, something will always be missing from our descriptions. Our measurements will never quite add up the way they ought to, and our laws will always have exceptions. For example, we won't be able to tell if a particular occurrence is explicable according to the design of the universe, or whether it results from divine tampering. Historians might study for centuries what caused some great war, and discover finally that their efforts were utterly wasted, because God started it on a divine whim. Likewise, it might be that the only reason gravity is so predictable is because God has so far desired it to be, and not because the physical design of the universe requires gravity to be predictable. Needless to say, in such an arbitrary universe, science is an exercise in futility. Our capacity for reason is wasted.

We must then assume that however our universe came into existence, it wasn't by divine creation. Creationists will continue to insist on the logical necessity of a first cause. But there needn't be a first cause. Nothingness, from which a zero-sum universe would spring, is utterly featureless. It has neither time nor space, and certainly no inherent characteristics like causality. Causality may be thought of as a *potentiality* in the cornucopia of Nothingness, but that's all. Causality is not an absolute. It only describes how events occur within this universe. Emergence from Nothingness requires no causation. Indeed, such an emergence can't have had a cause, because if it had a cause, it had an agent, which is impossible in Nothingness.

It must be noted that these reasonings don't directly challenge the existence of God. Our purpose is only to demonstrate that our universe is best viewed as objectively unreal—hence equivalent to nothing, and self-enclosed. Certainly God can exist within our universe. Not only is that idea not inconsistent with the paradigm we are formulating, it is encouraged by it, since in a zero-sum universe events will occur *as if* governed by a sentient god. Rather than God being a localized or centralized entity, however, God would be dispersed throughout the universe in its very design—omniscient and omnipresent.

5. A Universal Design

If we study the universe with the understanding that it has no objective existence and is actually a zero-sum, self-enclosed entity, we realize immediately that it can exist only subjectively. That is, the manifestation of our universe depends entirely on points-of-view within it. Not only could it not be perceived from outside, even if there were some point-of-view outside it, but it would not even exist for that outside observer. It is, however, quite real to any point-of-view within it—having substance and force that are undeniable.

Certainly the universe existed long before we came along. It existed for billions of years before sentient life could have managed to evolve or survive. There were no stars yet formed, no planets, none of the heavier elements. So, the question looms large—how could it have subjectively existed, if it had no subjectivity to sustain it? The answer is simply that it could not have. There can be no subjective universe without any subjectivity to sustain it. From this we must draw the startling conclusion that subjectivity is an inherent property in our universe. Not that it appears from inanimate matter only with the evolution of living beings, but rather that our universe begins as subjectivity realized. Our universe *is* subjective awareness, right down to hydrogen atoms, and very likely the protons and electrons within them. The modern view that some thinking ability is required for awareness to appear is incompatible with the understanding that a zero-sum universe can cohere only in the presence of subjectivity. Subjectivity is then more than just a mere feature of our universe; it is the essence of it. The forms that these points-of-view assume appear to us as matter, while changes in the points-of-view appear to us as energy transfers.

The apparent complexity of our universe hangs, then, on two fundamental principles. The first is that it begins as and always remains a zero-sum construction. The second is that all of its internal dynamics are arranged around one purpose—the creation and continuance of subjective points-of-view that make it real. Everything in our universe follows from these two precepts. That includes the emergence itself. When Somethingness emerges from Nothingness, it does so by a design specific to the purpose of creating subjective points-of-view.

The emergence is not a haphazard making of matter and energy that incidentally invents forms capable of subjective awareness. Rather, the emergence is a precise process organized for one and only one purpose—to create the subjectivity that imbues and realizes the universe.

If the casual reader assumes that the emergence refers only to the "big bang" itself, the significance of that last point is overlooked. To the Nothingness from which our universe springs, the notion of an absolute time-line along which events occur is absurd. Nothingness has no dimensions, thus no flow of time. To the Nothingness, the emergence is occurring at every point in our time-line. No point is any different from any other. This means that every event in every corner of our universe is governed entirely by the design of the emergence. The universe is continuously emerging by a single design or process. And, that design must rule every aspect of reality—the microcosm and the macrocosm, as well as both physical and abstract reality.

This interpretation of how our zero-sum universe is designed to function acts as a theoretical basis for the idea advanced at the outset of this paper, to wit that there is a single reality and that it can be studied as a thing-in-itself. We will shortly return to discussing how that universal design creates systems of symbolic form and directs how they change. Before doing so, however, we can conjecture a little further on the metaphysical structure of our universe.

6. The Topography of Space-time

The universe isn't really a vast spatial matrix through which time flows. Instead, it is an assemblage of many tiny spatial matrices, each with its own flow of time. These flows of time are not parallel to each other, although each is perpendicular to its own systemic spatial matrix. The further apart two systems are spatially, the more their temporal flows—like their spatial matrices—are tilted in relation to each other. Here is an analogy: if we scatter a handful of children's jacks on a sheet of paper and glue them down, and then bend the paper so that it curves parabolically, we'll see the matrices of the jacks tilt in relation to each other. Similarly, we can imagine our universe as a parabolic cone with the point at the top being the relevant point-of-view. Below it along the surface of the cone are all other systemic matrices (i.e. points-of-view). The further down the cone the other systems are, the more their matrices are tilted away from the point-of-view at the top. Time always flows at a right angle to the surface of the cone, so that every system perceives time as perpendicular to itself—but of course the flow for any one system is not parallel to the flow for any other system.

What is especially noteworthy is that the further down the cone one looks, the slower the relative rate of time appears to be. Furthermore, location on the cone determines not only spatial location, but also temporal location. That is, every system on the sides of the cone is behind the top point-of-view in time. If the point-of-view at the top is at 2007, a little further down the cone a system is still in 2006. Much further down, systems will be millions or billions of years behind. Because the rate of time is slower the lower one goes on the cone, systems below the top are falling further and further behind in time. A system on the side of the cone might be a century behind the point-of-view on the top, but a year later it might be a century plus one day behind. The overall effect is that of the top of the cone steadily rising in time, while the parabolic cone below stretches out. At the bottom of the cone are systems so extremely tilted in matrix orientation, in relation to the top, that they appear to be at the edge of the universe, and are falling behind in time so fast they appear to be traveling away at light speed.

Thus, when one maps the universe one is mapping not only spatial locations, but also temporal locations. The local system inevitably appears to be at the top of this cone, which it perceives as the center of the universe. And to it, the universe does not look like a cone. It looks, in fact, flat. This is because we—being always the point-of-view at the top—automatically flatten it out with our perception. As a result, it looks to us as if time falls upon the entire universe in parallel rays, as if distant galaxies are speeding away from us at the velocity of light, and as if the photons we receive from them have taken a long time to get to earth. But all that is part of the illusion caused by flattening the four-dimensional matrix down to a three-dimensional image. Time is not falling on everything in parallel rays; time that falls on any distant system really falls at an oblique angle to us. Distant galaxies are not speeding away; they are really falling further and further behind us in temporal location. Furthermore, the light we receive from them does not take time; it is immediate.

This last point bears further comment, because it so dramatically contradicts the current paradigm of physics. When we absorb a photon from a faraway star, it leaves the star at the same instant it hits us. It is only because the star is so far behind us in time that it appears as if the photon has needed light-years to reach us. In actuality, the speed of light is instantaneous.

Imagining the topography of our universe this way resolves several theoretical inconsistencies. In a context of instantaneous light, the results of delayed-choice experiments make new sense. And, if the angle of light falling on other systems is oblique to our frame of reference, making their rate of time slower, then it is easy to explain why we see relativistic distortions.

7. The Universe as a Singularity

At the foundation of western science is the belief that we live in a universe that is objectively real, but, as we have shown, that belief is fraught with logical difficulties. What is far more reasonable is that we live in a universe real only to subjective points-of-view built into it. And, that all form in this universe is made up of systemic subjectivity—that the design of this universe has a single purpose, which is to manufacture new points-of-view out of passing time.

Moreover, it seems probable that this not-objectively-real universe of ours is better described as a shared illusion. Every single system—from the atom to the human being to the spiraling galaxy—is at the center of *a* universe, and the overlay of all these countless universes is what we perceive as a single objective universe. There are an infinite number of such universes possible, of course, but each sees only those that are logically consistent with itself, and permit a collective whole that is equal to Nothingness.

It must also be kept in mind that Nothingness is dimensionless. This being the case, our universe must be a singularity, in which space and time are illusions. It isn't too hard to accept that time is an illusion, since we exist in a slice of it, with past and future as abstract constructs. But space is extensive in our universe; there is distance between things. If distance is an illusion, wouldn't all things be in the same spot? The key to answering this paradox is recalling that we exist only in a slice of time. Thus, time *can* separate things, while space, which is extensive, cannot. Odd as it may sound, time is space. Anything that appears to be apart from us in space is actually below us in time—that is, in an earlier time. We can picture this arrangement as all systems in the universe being stacked like discs along the temporal dimension, with the relevant point-of-view at the very top of the stack. When one looks further down the stack, one is looking back in time and seeing the discs beneath as if they are spread out in a three-dimensional space. The direction in which a lower disc appears probably depends on its internal balance of forces—its orientation in the four-dimensional matrix.

If that orientation is modified by an application of force, the disc appears to go into motion in the illusionary three-dimensional space. Its temporal location may then change, and if it does its rate of time will change as well. But even if no force modifies the orientation of the disc, it is steadily dropping behind the top disc in temporal location, due to the fact that its rate of time is slower. So, just as every disc sees itself as the center of the universe, so does it also perceive the universe expanding away from it. Which it is in fact doing, since temporal distance is also spatial distance.

We have employed different analogies and metaphors to describe our universe metaphysically. They have been presented as tools to help us visualize space-time, which can never be apprehended directly, since it is a representational illusion. If an aspect of our universe were seen four-dimensionally, it would look like an extremely complex Mobius strip. We see it flattened into a three-dimensional image. An analogy for this would be flattening a paper Mobius strip down into a two-dimensional representation of itself. When we flatten space-time down into three dimensions, the four-dimensional Mobius curves are flattened into form, with their constituent periodic motions functioning as a record of time passed. And as time continues to expand that form grows continuously deeper and more complex.

Whether we imagine the universe as a parabolic cone of spacetime, or a stack of system-discs, or a complex Mobius strip, we must remember that it is actually a dimensionless singularity, seen representationally. Could we see the entirety of it rather than just the representation produced by collapsing time down to a single slice, we would see Nothingness, which is all that is really there.

* * *

8. Systems of Periodic Motions

The best way to understand reality is not as an array of things that cause change and are caused to change, but rather as a world of interactive systems. In this world, in fact, there is nothing but systems. All things are meaningful elements in one system or another, and these systems overlap in many ways often unnoticed. A man in a factory, for example, is himself a system. He is also a part of the factory-system. The factory-system is part of the industry-system—be it the furniture industry or the steel industry or the widget industry. The industry-system is part of the economic system, which itself is part of the nation-system, which in its turn is part of that great global system we call humankind. An atom is a system, but so also is a tree, and a forest, and a tide-pool. We have what we call ecosystems, weather systems, political systems, mechanical systems. Everything that exists in reality is part of some system. And what is just as important, nothing has any meaning except insofar as it is part of a system. Meaning is always contextual.

Of course, we don't go through life interpreting the world only in terms of systems. When one reads a book one cares not how it fits into the cultural system, but instead if it engages one's interest. In the same way, an automobile may belong in the national transportation system but its owner sees it in more personal terms. Also, because of the complicated relations between systems, they can't be readily sorted out, with each one delineated as distinct from others. Systems tend to be somewhat indeterminate. Remove a frog from its pond, and who notices its absence? The ecosystem of the pond is unchanged. Burn a pile of leaves, and what discernible effect does the particulate smoke have on the global weather system? None, so far as we can tell. Systems, for the most part, have plasticity and robustness. They did, after all, evolve to survive under a range of conditions. Thus we can go about our lives viewing things as just things, and concern ourselves very little with how the things fit into larger systems.

Nevertheless, seeing reality in terms of systems is the key to understanding it, because systems all have the same basic structure. According to Webster's, a system is "a set or arrangement of things so

29

related or connected as to form a unity, or organic whole," but the unity or whole is not static. For example, a factory observed from afar does not reveal the activity within. And, the internal motions are far from random. They follow certain oscillations and rhythms and cycles. The periodic motions are what make up the unity, in fact. If internal rhythms cease, the system ceases. When workers in a factory shut down the machinery and go home, what remains is only a factory building. Likewise with the system that is a living being. When the heart stops its rhythm, a rhythm that is crucial to all the other cycles and rhythms in the body and brain, the living being ceases to be a system and becomes a corpse. Even the humble atom, from which all things are made, exists only because of its internal motions. It is composed of speeding bits of energy; what gives the atom its constancy and substance is that the quanta curve around each other, locked in rhythms and cycles. When an atom is "smashed," these interlocked motions are disrupted. Energy shoots off every which way, and the atom ceases to be an atomic system. Any system is defined by its internal motions, since these are what give it continuity. The motions are pulsing, oscillatory, wave-like, and cyclic in character.

It is plainly evident that the pulses, oscillations, waves, and cycles are periodic motions of energy. In fact, systems themselves are nothing but arrangements of energy. Even the matter in a system is actually energy bound up in systems we call atoms, and the atoms can be converted directly to energy, according to the equation $E=mc^2$, by disrupting the periodic motions in atomic systems, thereby freeing the energy to disperse randomly. So when we talk about how systems behave, or interact with each other, we are necessarily talking about energy—how it is bound up in systems and exchanged from system to system.

9. Energy

It may come as some surprise to laymen that the term *energy* has never been defined except as an expression of what it can do. Energy is equated with power and force and is measured by its capacity to do work—i.e., to overcome resistance. Which is to say that even the best scientists don't know what energy really is. They only predict how it will behave.

According to physics, all energy in the universe results from four fundamental forces: the strong force that binds atomic nucleons together, a weak force associated with beta decay (and, it seems, with the property known as *spin*), the electromagnetic force that underlies magnetism and light, and the gravitational force. These fundamental forces are responsible for all energy in the world. They give rise to all material things as well as to whatever energies those things possess. In these pages, however, we go a step beyond the standard paradigm and assert that the dimensionality of the universe is energetic in character. Thus, not only is space-time permeated with fields of energy, it *is* those fields, and without them it would disappear. Where there is energy there is space-time, and where there is no energy there is no space-time. The reason for this intimate connection will become evident shortly.

We can describe our universe like this: everything we perceive is made of knots of energy, which we call atoms, while all events involve flows and exchanges of energy, and the background against which things exist and events happen is a four-dimensional construct of attenuated energy fields.

Yet there is an important point to make here, which is that energy does not exist in and of itself. Rather, it is the tendency of objects to behave in certain ways. It was Einstein himself who made this distinction when he pointed out that there is no action-at-a-distance between two masses. The term *gravity* describes an effect of the distortions in space-time. There is no mysterious force that pulls objects together. It is the curves of space-time that bring them closer. In this Einsteinian universe, what we call the gravitational force is revealed as an illusion. The same is true of the fields of the other three fundamental forces—they also are but distortions in the fabric of space-time. Objects may be made of energy, but energy is only the warp and woof of space-time.

Einstein's insight was revolutionary and is conceptually difficult to grasp even today, but it is an accurate description of our universe. What we call reality is a tapestry of interwoven fields, fields that are just distortions in space-time. Meanwhile, space-time itself is a zero-sum construct. What we perceive as a universe is the twisting and curving of that zero-sum construct. The objects we see are stable knot-like complexes of twists and curves—i.e. mass. Other curvatures are simpler distortions, and we call them energy fields. When change occurs, this woven substrate is slightly modified, with retwisting and reshaping of its constituent distortions.

For most purposes, it is easier to understand the twists and curlicues of this space-time substrate as being made of atoms and energy fields superimposed on a background of empty space. Having been around since at least the time of Democritus, that is a familiar interpretation of reality and deceptively easy to visualize. It is also, unfortunately, an inaccurate description of microcosmic reality, and even serious physicists have been seduced by its facility. We can use it to formulate laws that predict how macrocosmic change occurs, but that isn't the same thing as understanding reality. As long as we consider reality in terms of mass, energy, and space-time, and search for a set of equations that relates them, the interface between Somethingness and Nothingness will elude us. That interface can only be understood when we see the universe as a zero-sum construct in which mass and energy express the dimensional substrate.

10. Dimensional Formats

As everyone knows, space-time has four dimensions, three spatial and one temporal. In this four-dimensioned space-time, it seems as if field energy manifests in just three dimensions. That is, a field appears to exist in space but not time. Imagine a magnetic field: it obviously has three spatial dimensions. Because it doesn't move or change it appears to lack the temporal dimension. Not true. The magnetic field, like every other energy field, does have a temporal dimension. It reveals itself as the vector of the field—the directional component of its force—rather than as outright motion. If the spatial dimensionality of the field is compressed, however, that static temporal dimension will manifest as motion. As an example, when part of the energy in an electrical field is transmitted as light its vector ceases to be a static property and becomes motion. The dimension of time in the otherwise inert electromagnetic field is liberated. We can picture this with an analogy: holding a stretched-out rubberband on the fingertips and releasing one end, so that it shoots across the room. The rubberband is still the same rubberband, but now its vector value—its tension—has been converted into motion.

Nor is that the only modification of dimensionality that can be performed on a field of energy. If one compresses that same three-dimensional rubberband down until it is completely flat, it will have two spatial dimensions. The third spatial will show itself as vibration of the rubberband, while its temporal dimension will show itself as motion. If one then squashes the rubberband sideways, it will be reduced to one vibrating and moving spatial dimension. Finally, if one squashes down its length with great force, it becomes a single rubber point and nothing but its temporal dimension will show—as a pulsing dimensionless particle. As observers we perform operations like these on fields of energy all the time, seldom noticing what we are doing. At any given moment each of us is manipulating trillions of dimensional formats in our personal corner of space-time—with physical motion, and also within the cells of our bodies.

Think about the electromagnetic field mentioned above. In its natural state, it has the three spatial dimensions and a static temporal dimension. When its spatial dimensionality is reduced it surrenders a quantum of energy, which is released as a photon. That photon leaves the field, propelled by its transformed vector energy, traveling at light speed. Its follows a one-dimensional trajectory and vibrates as it travels. The vibration expresses itself in three or two dimensions, as a corkscrewed helix or as a wave, depending on how flattened the vibration is. That is, it has either an elliptical or a plane polarization. Depending on how many spatial dimensions we measure it in, this photon will exhibit either wave-like or point-like properties.

Electromagnetic energy is often in motion, exchanged between systems as heat and light. Except for gravity—which is quite easily distinguished because of its constant nature—almost everything we notice is some kind of electromagnetic energy exchange mediated by dimensionally-collapsed waves and particles. The energy fields of the strong and weak forces we don't normally perceive in our everyday world, but quantum physicists have found ways to collapse their spatial dimensions into a variety of moving and vibrating subatomic particles in the laboratory. They've had no luck producing gravitons, which would be the gravitational equivalent of photons. And, efforts to detect gravity waves have been unsuccessful thus far.

It should be noted that the natural state of a field is full four-dimensionality. Although a field can have its full dimensional appearance reduced, so that spatial dimensions express as vibrations and the temporal dimension as motion of the spatially-collapsed field, it reverts to four-dimensionality as soon as possible.

11. Representing Reality

Energy fields are like everything else in our world: they have different appearances depending on how many dimensions are used to represent them. We can explore this idea using a football game for an analogy. The game is played out in four dimensions—width, length, height, and time. For the fan watching from his stadium seat, the game plays out on his retinas in just three dimensions—width, height, and time. That is, he sees an image in two spatial dimensions changing over time, just as he watches the game on TV. His brain infers a third spatial dimension, giving the image depth—a task made easy by considering perspective and easier because his eyes give him binocular vision. Now, let us say the person uses a still camera to take photos of the game. Then he sees the game in two dimensions, since a photograph is a static image on a flat piece of paper. Suppose that he uses a digital movie camera to film the game. Then, the game could conceivably be watched as a stream of bits (although no one would want to watch it this way), having been reduced to just two dimensions—one spatial dimension, plus time. We can further reduce the game to just one spatial dimension, by putting the final score on a bar graph. Finally, we can collapse the dimensionality of the game down to a dimensionless point. Imagine that the team's season is shown in a chart or table where a dot indicates a won game and no dot a lost one. Now the game, originally played in four dimensions, is represented in zero spatial dimensions. It still has a temporal dimension, since on/off periodicity—like in a computer's memory bit—is a fundamental motion.

What we begin to realize, as we study this phenomenon, is that perception *is* representation. Any act of perception—or observation, which is the generic term for perception in physics—reduces the perceived object or scene in number of dimensions. This is true in both the microcosm—as in the subatomic realm—and in the macrocosm of our everyday reality. Representation (which literally means re-presenting) always involves reducing the number of dimensions that the original field possesses. We represent an electromagnetic field by collapsing its spatial dimensions into a wave or particle, just as we represent a real-life event—in perception, communication, or portrayal—by collapsing its dimensions. Because our reality is real in only a subjective sense, it depends on observations to sustain it, and all such observations involve the dimensional collapses typical of the representing process. Again we must remember that, directly or indirectly, all these collapsed representations return to the four-dimensional format as soon as possible. It is fair to say that our reality evolves through these equal and countervailing processes: the collapsing of dimensions with observation, and the restoring of full dimensionality to that which has been collapsed.

When we spoke of the universal design by which Somethingness emerges from Nothingness, this is what we referred to, because the universe itself is—like all its contents—representational. How else could it exist? The only way to construct a universe from Nothingness is by not allowing the entire Nothingness to be perceived. If it were perceived in its entirety, nothing at all would be there. However, representations of Nothingness *can* be perceived. The universe with which we are familiar can emerge from a Nothingness if that Nothingness is subdivided four-dimensionally, then represented three-dimensionally. The representation will always zero-sum, but to internal points-of-view it will appear as if it were objectively real. As we demonstrated earlier, however, it isn't. Physical reality is just a representation of Nothingness, and the universal design that governs that physical reality is one that generates internal points-of-view essential for it to subjectively exist.

According to the paradigm outlined here the universe is a dimensionally-collapsed representation, its contents are representations as well, and these contents collectively sum to zero. The next question to be addressed is: why is form so ubiquitous in our universe? The current paradigm of physics has not been able to offer a plausible answer, but our new paradigm can.

12. Time, Space, and Form

Perhaps the most baffling aspect of our universe is the eagerness with which it organizes into forms. This tendency is so pervasive and so natural that it even challenges the notion of a divine creation. It is easy to insist that God made everything, but new things constantly appear, in raw nature as well as in technosocial spheres of activity. Were these just latent in the divine creation? Or is the creation an ongoing project for God? It is one thing to ascribe the existence of atoms and stars and planets to God's creativity, but what about the Internet? What about bioengineered viruses? What about each new and unique hurricane? What about an evolving fungus? What about calculus? If God invented calculus in the Beginning, where did it hide for billions of years until Newton decided it was needed? These questions aren't offered as proof there wasn't a divine creation. Instead, they are intended to illustrate the fact that whenever and wherever free or chaotic energy appears in our universe, it quickly organizes into form. Nothing ever remains disorganized. Self-organization is the name of the game. To understand why, we must reflect on how representations are made.

To be a true representation, an image or likeness must allow for all of the dimensional values of the original it purports to represent. For example, when an artist represents a scene on his canvas he skillfully uses shading and perspective, so that a third spatial dimension is implied. In the mural paintings of Ancient Egypt this wasn't done, which is why those images look quite literally flat. But it isn't sufficient to indicate just the three spatial dimensions in a piece of art, not if it is to be an accurate rendering. The dimension of time must be accounted for. In a good work of art, the temporal dimension is suggested by such a skillful arrangement of the image that the eye is guided in a circular or elliptical pattern of scanning. The painting or a sculpture doesn't permit a static gaze, keeping the eye instead in continuous motion. This makes the work look real—i.e., four-dimensional.

Representations always reduce the number of dimensions in which the original is formatted. But the dimensions eliminated must still be implied if the representation is to be reasonably accurate. Time is always eliminated in painting and sculpture, but that dimension will always be implied if the work is successful. Spatial dimensions can also be eliminated in a representation. When the earth is mapped onto a two-dimensional surface, its third spatial dimension is eliminated. That spatial dimension insinuates itself into the map as Mercator-type distortions. Similarly, when a cardboard carton is crushed flat, the sides are bent and distorted. In the latter examples the distortions are residual evidences of the spatial dimension that was eliminated. The distortions are the equivalent of the dark shading and perspective adjustments of angle and relative size that indicate the elimination of dimensional depth in photographs and paintings.

With our physical universe, it is also insufficient for the representation to indicate only the values of three dimensions. Although the fourth dimension can't be rendered explicitly, it can be implied, and it is, through the curvatures of the dimensional substrate. These are the equivalent of shadings and perspective in the representation that is our universe. Time, the eliminated fourth dimension, shapes everything in our universe—whether matter or energy, physical or non-physical—into distortional curves.

The curves ultimately become cycles. Cycles, of course, can be dimensionally reduced to waves and oscillations and pulses, and wherever we look in our universe we perceive it to be entirely made up of such periodic motions. These are the residual evidences of the eliminated fourth dimension that we call time. As the spatial dimensions of the universe expand, the temporal dimension does also, and as it is continuously integrated into the representation—maintaining its representational accuracy—the coils of cycles gradually accumulate. Form evolves. It must, if our universe is to remain an accurate three-dimensional representation as its dimensions expand.

13. Dimensional Collapses and Time

Systems, as we previously noted, are composed of periodic motions—cycles, waves, oscillations, and on/offs. These periodic motions are energetic. To test that statement, one can insert a fingertip into the blades of a whirling fan, or touch a vibrating tuning fork to the tongue. However, although every system is full of energy (and is made of nothing *but* energy), that energy is locked up in fields that don't display any visible motion. Consider the fan: if the rpm's of its blades are high enough, even such obvious motion is not visible. Something similar happens with the electron which "circles" about the atomic nucleus; it turns into a blur more like a static cloud than a revolving particle. Now, the layman is inclined to believe that no, the electron isn't really a cloud—it looks like one because it has such rapid motion. Not true. The electron is in fact spread out around the nucleus and doesn't have any specific location at a particular time. Our explanation here becomes controversial because we aren't just talking about electrons. Most readers will remember Zeno's paradox of the flying arrow, which asks, if the arrow is always at some particular spot in its trajectory, when does it move? And we are now answering that paradox by saying that the arrow is in one particular spot only when it is observed. Otherwise it is, like the electron, a blur. Of course there is no easy way to test this view, because to test it one must observe, and by observing collapse the arrow into specific space. And, in our macrocosmic world so many observations are occurring constantly, everywhere, that the point is moot. Every molecule of air the arrow strikes observes its passage, even if there isn't any person watching and following the trajectory of the arrow. But this interpretation makes possible an even more startling realization about systems, to wit: in the ideal system, unobserved, there is no passage of time. The system is composed entirely of cycles, which manifest as static energy fields, so there is no internal motion at all. Without internal motion, there is no internal passage of time. If an observation is performed on the system (such as poking a fingertip into the fan blades, or firing a neutron into the electron cloud), the static four-dimensional fields partially collapse, and motion of some kind occurs; hence time passes. From this interpretation we discover a most useful key to understanding reality. Time passes because of—and is mediated by—observations.

Again, we are speaking of ideal systems, and these exist only in the microcosmic realm. For an atom drifting alone in deep space, there is no time. However, if a photon of starlight happens to strike its electron clouds, it observes that photon, and the atom advances in time. In our macrocosmic world all systems are imperfect, and are constantly subject to veritable storms of observations—not only on them, but also by them. Their fields are continuously being collapsed, so that there is never a moment when time actually stops passing for them. Nevertheless, when we look at many real-world systems, particularly biological ones that are adapted to the endless barrage of quantum-type observations, we notice that in an abstract sense time does stop for them when observations of the macrocosmic type cease. For a person who is extremely bored, for example, time seems to dramatically slow down. Or, let us consider those primitive cultures where ritual and tradition (as well as isolation) constitute powerful protection against observations of self and society and way of life. In these time also stops—for the culture, if not the individual—and the culture may show no change for thousands of years. Should a dramatic observation be made, however—such as the discovery of writing, or the invention of the printing press, or development of Freudian theory—time and change accelerate.

Our universe, as we have described it, is composed of systems. But the dimension of time is fully immersed in those systems. Time doesn't flow forward in a system that is isolated from interaction with other systems. Only when an observation takes place is part of the four-dimensional matrix of a system collapsed, and the temporal dimension freed. Only then can time pass and change occur.

14. Observations as Energy Transfers

What does the word *change* mean? What is it that happens when an observation induces change? We see constant change all about us in everyday reality. Things grow or decay, things get better or worse, things become different than they were, so there seem to be various kinds of change. But all changes have one characteristic in common: they involve transfers of energy. Whenever an observation takes place, energy is transferred from one system to another. No events occur at all unless dimensions are collapsed by observation. When dimensions are collapsed and time is set in motion, energy moves to the system doing the observing (which we will call S1), from the system being observed (S2). S1 thus has a net energy increase, while S2 suffers a net energy loss. So, a change occurs when an observation by S1 causes a dimensional value in S2 to collapse, thus setting it into motion, ultimately to be absorbed by S1.

There are three general categories of energy transfer. In the first, S1 takes from S2 and gives S2 nothing back. Visual observation is an instance of this category. When S1 (in this case the retina of the eye) observes S2, it takes from it a photon of electromagnetic energy. In most of our visual observations what we actually observe are solar photons reflected off a local surface, so S2 is the sun itself and it suffers a net energy loss. With firelight or incandescent light bulbs, the eye captures photons released from heated sources. Another example of this type of transfer is heat exchange. The vibratory energy of atoms in S2 is absorbed by atoms in S1 as respective temperatures equalize. Conventional ovens use this observing process to heat food.

In a second category of observations are the energy transfers that work in both directions: S1 and S2 surrender energy to each other and receive energy from each other. Whenever two systems touch each other or collide there is a mutual exchange of energy of this kind. Even if the amounts exchanged are perfectly equal, both systems are dramatically affected. If a person strikes two dinner plates together like cymbals or rolls two billiard balls into each other, the results demonstrate that equal transfers do not just cancel out.

A third category includes those events where S1 enters into a force-field of S2—e.g. gravitational or magnetic. This type is especially interesting, because the only immediate evidence that S1 has gained energy is the increase in its velocity (and thus relativistic mass). Consider the bomb that is dropped from a plane: its kinetic energy content rises geometrically as it falls, although its rest mass remains the same. When it crashes to earth, that kinetic energy is released as pressure and heat, and the explosive is detonated. But where did the kinetic energy originally come from? There can only be one answer—from the field itself. The bomb (as S1) is collapsing the gravitational field energy of the earth (S2) as it falls, reducing that four-dimensional field to a temporal dimension. The earth is simultaneously observing the bomb in the same way, of course, but due to the difference in mass the effect is negligible, and we don't notice the earth rushing up to meet the bomb. Such field energy exchanges are distinct from the second category because they have a delayed effect. The observations produce no visible changes in the involved systems unless or until collision occurs.

We can see from this analysis that the concepts of energy, observation, time, system, and change must be understood in a package. Our universe is composed of representational four-dimensional systems, and if these systems are undisturbed by observations they remain static. Time does not pass and nothing changes. With observations, some dimensional values are collapsed and exchanged as quanta of energy, allowing time to flow as the changes occur. These dynamics of change are exactly the same in the everyday world as they are in the quantum realm of energy exchanges.

* * *

15.   The Energy Needs of Systems

Systems have energy needs. Without inputs of energy, a system will sooner or later collapse. There are two reasons why. The first is that a system is a living entity, animated by the flow of energy through it. When the input flow stops, the system becomes dormant or dies. Biological systems have stores of saved energy to guard against sudden loss of outside energy supplies, and non-biological systems can often shut down temporarily, but generally speaking every system is like a hot air balloon—as soon as the flame is extinguished, the balloon begins to collapse. The attentive reader might raise a question about atoms, because we have classified them as systems, yet they seem to cohere without an input energy source. However, it is possible that atoms and their constituent subatomic particles are sustained by the universal expansion of the dimensions. We are inclined to hypothesize that should that expansion cease, time itself would stop, all oscillatory and wave motions at the atomic level would end, and all the universe would instantly collapse back into a singularity. Needless to say, this is not a hypothesis we would want to test, even if we had the capability. In any case, it is at least true that systems in the everyday world do require inputs of energy to cohere even if atoms do not.

The second reason that a system needs continuous inputs of energy is that it is always losing bits of energy to other systems that are observing it. As we explained before, an observation involves an energy transfer from the observed system to the observing one. There can be no observing without taking. To observe, one must collapse part of the dimensional matrix of the observed system, and when that happens the bit of collapsed dimension migrates instantly to the observing system, then is integrated into its dimensional matrices. Although it is fairly easy to recognize such exchanges at microcosmic levels, the most familiar case being the transfer of a photon from seen to the seer, it isn't always as obvious in real-world events. Let us offer the straightforward example of being hit in the head with a stone. The person who is hit is observing the stone and absorbing much of its energy in doing so. How did the rock get the energy it gave up to the head? By observing the hand of the person who threw it. Where did that person get the energy to throw it? By digesting food, which is the process of observing energy-rich molecules. The food, if from plants, got its energy through photosynthesis—that is, observing sunlight. This chain of energy exchanges is also a traceable series of observations. If there is any confusion as to which system is observing and which is observed, one need only determine which system is gaining energy or which losing it.

We might also comment here on energy transfers involving more abstracted systems, as it is impossible to understand the dynamics of social, psychological, political, and economic change without understanding that such exchanges do occur, even when not noticed. As an example, the IRS is observing a taxpayer when it receives a tax return, and it hits the energy jackpot if that return includes a payment. The demagogue politician gains power/energy by denouncing—which means making an observation of—a target enemy. The teacher collecting test papers from students is observing them, and in doing so taking energy from them. In these examples, the energy lost or gained can be tangible or intangible. It may well be monetary, as it is with taxes, or it may be psychological, as it is with the demagogue rhetorically defining his target group. It may be pure physical energy measurable in calories, as is the energy surrendered by students taking the test, while the gain to the teacher is an informational one. Real world transactions can take many forms, but they are always exchanges of energy through observation.

If any system is to survive, it requires inputs of energy, not only to sustain its manifestation, but also to replenish the energy it loses to other systems that are observing it. These inputs of energy will come from observations that affect the systems that it observes in two possible ways.

16. Analytic and Synthetic Processes

In this world of interwoven and interactive systems that we have been describing, all systems need outside sources of energy. There is no such thing as free energy, because all energy in the universe is bound up in systems, so when any system seeks an outside source of energy it must look to other systems. It has to collapse part of another system's dimensional structure—by the process of observation—and absorb energy that is released in that collapse. There is no other way for a system to procure energy. Whether one is burning coal, collecting solar power, digesting food, levying taxes, heating geothermally, assigning tests, stealing from the poor box, taking donations, invading another country, or pillaging a town, one is collapsing the dimensional structure of other systems by observation, and profiting energy-wise in doing so.

However, that having been said, one of the examples is intrinsically different. While the others have an analytic effect on the provider system, the collection of solar power has a synthetic effect on the sun. Hydrogen atoms on the sun's surface, when observed, fuse into heavier helium atoms. Binding energy is released, because one helium atom needs less energy to cohere than the two hydrogen atoms that fuse together do. Yet helium is a more complex form than hydrogen, so we arrive at the counterintuitive realization that taking energy from the sun by observing it is a constructive process.

At first, this process seems anomalous. All the familiar methods of getting energy for our needs are destructive, in the sense that they break down complex forms into simpler forms. Oil burned becomes particulate smoke and wood becomes ash. Collecting taxes leaves citizens poorer, digesting foods removes energy and nutrients, testing students leaves them exhausted, and so on. These are what we can call analytic processes; they analyze the energy source into simpler parts. But there is an equally powerful method for producing energy, one that we can call synthetic, since it acts by the same kind of fusion as does the sun. Far from being anomalous, synthesis is ubiquitous in our world. It is, in fact, the only reason that form can evolve into higher forms at all: if there were no energy savings in the synergy of cooperation, the only life on this planet would still be microbial.

A system can be imagined as an interwoven four-dimensional latticework. In this structure, each discrete section is a bit of dimensional value. When an observation is performed on this structure, some bit or another collapses and is captured by the observing system. In an analytic process of observation—such as burning wood for heat—the structure is progressively robbed of dimensional components until the systemic structure itself collapses. In the synthetic process, discrete sections of structure are overlaid on other discrete sections, creating a new form and in doing so freeing up unneeded bits of dimensional values. Sometimes, the synthetic process shortens an unnecessarily long section of the lattice—either a spatial or temporal bit—without compromising the structural integrity of the system.

Using the synthetic process to meet energy needs is far more effective than using the analytic one. Since systems require a constant energy input to sustain themselves, reduction in the amount of necessary structural components lowers the necessary energy input permanently, besides offering the one-time shot of energy. On the other hand, the analytic process only provides a one-time release, and even this gain must be weighed against the damage done to the source, and cost of clean-up.

Although we have cast this contrast in terms of energy production from fuels, the same principles apply in any situation where observations are employed to harvest abstract types of energy.

17. Change in Systems

The evolution of form in our universe progresses by means of observations. Observations are the growing pains of the universe, so to speak. While in the short run observations look only like tiny exchanges of energy, over the longer course of events they become new symbolic form. We should all be familiar with how this transformation happens. A person notices some event, then a similar event and yet another, and then suddenly these observations turn into a generalization. All of reality works the same way. Cumulative observations become generalized symbols, which are the material of symbolic structure. The birth of galaxies and solar systems, the process of genetic speciation, the development of cultures and canons, the growth of towns and cities—these are all examples of symbolic form evolving through cumulative observations. It is with this process that the universe converts the expanding values of space and time into three-dimensional representations, and it is the process that governs change in our everyday reality.

Where there are no observations there are no changes, and time stands still. Where observations do take place, energy is exchanged between systems, and both the observed and the observing system are affected by change. It might seem that, with such a diverse variety of changes in our world, little more could be said on this subject. However, the effects of energy exchanges are to a surprising extent predictable. We can discuss these briefly.

Firstly, the loss of energy for the observed system means that it must internally re-balance itself if it is to hold together. It has lost a part of its dimensional structure and now it must re-arrange its remaining energies to compensate. This can be as simple as an atom dropping an electron into a lower shell, or it can be as dramatic as a defeated empire learning to live without its colonies. Clearly, the energy loss can be so severe that re-balancing leaves the system substantially weakened and less efficient in how it uses input energy. This is not necessarily the case, however, because the loss can also lead to internal syntheses that leave the system stronger than before.

Secondly, the gain of energy for the observing system means that it must either find a way to pass it through or else somehow absorb it. If the energy has a single vector, as when a baseball is whacked by a bat, and the ball can spread the force through itself evenly, then it is only re-oriented in the four-dimensional space-time matrix and its momentum is changed; it then flies through the air. If the energy has multiple vectors, however, such as when a stick of firewood is heated, at first the energy will be distributed through its structure, but with continued heating the cellular systems will begin to disintegrate. Parenthetically, this is an example of how living creatures harvest energy from other systems—by forcing a barrage of observations on the target system until it can absorb no more and begins to chemically break down, releasing huge amounts of energy, which are then captured.

Overall, an energy transfer can cause considerable difficulties for the system that receives the energy, as well as the one that loses it. It is a dangerous world out there, for the observers as well as the observed. Basically, there are three general risks for a system. Its input energy flow may diminish, in which case the system tends to stagnate, with differentiation vanishing and change slowing down. Or, differentiation may proliferate in a system until the input energy can't support it, in which event the system usually separates into discrete functioning mini-systems. The third risk is that a system may be so overwhelmed by a barrage of energy that its internal balances crumble; in this case the system is typically destroyed, with its debris scavenged and absorbed by contingent systems.

The first of these categories results from too little input energy and the third from too much. It is the second that is most interesting, because it has been so poorly understood. We turn to it now.

18. Differentiation

Ideally, a system would always have input energy sufficient to sustain it. The real world seldom works that way; multiple variables affect energy flows, causing fluctuations. This is especially problematic for biological systems, which are perched atop the non-biological systems of nature, therefore vulnerable in myriad ways. Drought and flood, climate change, volcanic activity, earthquake, and meteor impact can all severely disrupt the flows of available energy that living systems need to survive. It isn't surprising that the history of life on earth is punctuated with mass extinctions. But when a local or global disruption ends, there follows a resurgence of biological systems. Soon newly differentiated species again capture the flows of energy. As long as there is sufficient available energy some life-form will appear to capitalize on it.

This universal tendency toward increased differentiation applies also to every individual system. If the input energy flow to a system increases, the system responds by differentiating. What else is a system to do with the surplus energy flowing into it? Systems are designed to maximize their input flows, not restrict them. Nor are systems designed so that energy can pass through them unused. Rather, they are designed as the universe itself is—to spend surplus energy on differentiations, to create and evolve subjective points-of-view. Thus, a herd in a lush environment grows rapidly in population, an overfed animal gets fat, and a well-fertilized tree sends out new branches. Whenever biological systems get more energy, they grow.

Human beings do the same. With people, however, surplus energy is often converted to abstract differentiations. In fact, we can state as a rule that when any human system has surplus energy, it invariably turns much of that surplus to the creation and manipulation of symbols. Family, tribe, nation, or individual, it always translates itself—its history and identity—into extrinsic symbolic mediums: art, statuary, temples, rituals, adornments, and entertainments. Unless the society with surplus energy deliberately restrains itself, it will also differentiate into classes that are based on division of labor but inevitably reflect access to and control of the surplus energy. Ultimately, it will differentiate its governing structure into what we know as bureaucracy.

For biological systems, the negative consequences of this natural tendency toward differentiation can hardly be overstated. Often the surplus of energy is just a temporary effect of a fluctuating variable, and when it returns to the mean these new differentiations are unsustainable. Also, it is rare that a biological system quits while it is ahead. Given opportunity, deer will overpopulate until mass starvation causes a die-off, a virus will multiply until it kills its host, and a bureaucracy will grow until its weight crushes those that it is supposed to serve. For human society, the most significant consideration is that any surplus flows soon cease to be surpluses. They are completely converted to differentiations—dispersed throughout the system in the same way that the impact force of a bat on a baseball is dispersed throughout the ball. Even if the surplus is the result of a technological breakthrough, and is a permanent improvement, it inevitably vanishes into the systemic structure. Profit is always a temporary phenomenon. It naturally diminishes, no matter how strenuously the profiteer guards it. Eventually it is no longer profit, but merely another requisite input of energy for the system. Meanwhile, the differentiations it funded will need routine maintenance. Where is the energy to come from? Children must be fed and clothed and educated and found productive jobs. New infrastructure wears out and needs repair. An expanded bureaucracy brings the expenses of medical care, vacations, pensions and the training of replacements for retirees. The surplus has been converted to differentiations with

little regard for these expenses. Sooner or later, the society that believed it was rich discovers that it can't pay its energy bills. Unfortunately, the methods it uses to address the problem often only
aggravate it.

19. The Vanishing Surplus

What happens to a system after an apparent energy surplus has
been entirely spent on systemic differentiation? This is an interesting
question, because the answer explains much of human history. It
must be said that for all the enormous work that goes into building a
civilization like our own, surprisingly little effort has been expended
on understanding its ontogenetic dynamics. Small wonder, then, that
western civilization seems to be repeating the rise and fall cycle demonstrated by every prior civilization.

All civilizations great and small begin as just primitive cultures.
Typically, the burst of initial growth for a civilization is driven by a
newly discovered source of energy. Usually this is a food surplus, or
technique for producing a food surplus, but the energy is sometimes
gained through synthesis, as it was with the Islamic Empire. Ordinar-
ily, primitive cultures develop rituals and traditions that preclude
growth and change, but a new energy surplus, by increasing population, can trigger sudden symbolic development—almost always with
writing of some kind. The sociocultural system then is subject to a
veritable barrage of observations and plunged into rapid change.
Time, which had stood still, is set into motion by the observations and
begins to flow. With the surplus being enthusiastically spent on differentiations, a period of remarkable achievement begins. What once
was a primitive culture is quickly transformed into a blossoming civilization.

Sooner or later the original energy surplus dwindles. As we indicated earlier, even if the flow itself continues, it all now goes to sustaining the new differentiations. It isn't just the burgeoning
bureaucracy that consumes the energy. Differentiations can also be
improvements in standard of living—in housing, clothing, and diet.

Improvements like better sanitation, transportation, and health care can generate a secondary wave of surplus energy. In the end, however, the surpluses all are accounted for by the extensive differentiations of a highly evolved civilization.

By this point the civilization has already overshot its mark. Even a general consensus of opinion that the surplus is gone won't solve the problem, because the surplus is actually over-committed. Too many of the differentiations have delayed costs built into them, and these unfunded expenses begin to bite into the economic pie. Furthermore, the expectation of a constantly rising standard of living has firmly embedded itself in the culture, and whatever elements of the population have political or economic leverage go on demanding more of a surplus that no longer exists.

Since voluntarily reversing differentiations is almost impossible, the civilization needs to find new surpluses of energy to support its differentiated structures. Some ill-advised but predictable responses attempt to deceive the system into believing it still has a surplus. One of these is to debase the currency; another is to inflate the money supply; still another is to borrow money from foreigners. Another and especially insidious solution is to sell off functioning parts of the civilization-system—capital infrastructure and jobs—to foreigners. At best these solutions only postpone the problem and at worst they hasten it. They provide only an illusion of new energy flows. There are but two real methods of producing energy—by analytic or synthetic process—and they use neither.

Sometimes a civilization is fortunate enough to discover a new outside source of energy. It might be something as crude as piracy and plunder. Stealing valuables, commodities, or slaves from other peoples is using the analytic process to generate a surplus. Using a new fuel, such as peat or coal or oil, does the same. The problem with the analytic process, as we noted earlier, is that it offers only one-shot infusions of energy. Far better are synthetic solutions like cultivating new trading partners and markets, signing peace treaties, and exchanging technologies. In any event, however, new outside sources of energy are generally too few and far between to sustain a continuously differentiating civilization.

20. After a Surplus Vanishes

It is easy enough to understand how an evolving civilization employs analytic and synthetic processes to capture the energy it uses for its differentiations. In most historical instances the new source of energy was a food surplus, which can be gotten by analytic means, such as hunting or fishing, or a more synthetic means, such as agriculture. Some civilizations, like the Phoenician and the British, were funded, because of their geographical location, by trade, which is synthetic in nature. The Mongols and the Vikings funded their differentiations through the analytic means of plunder—made possible by their horses and their ships, respectively. The growth of the early American colonies was initially funded analytically by its incredibly rich natural resources and later through the syntheses of manufacturing. In every case, whether the surplus arose from analytic or synthetic means, the energy gained was spent on differentiations like population growth, improved housing, roads and cities, schools, armies, the bureaucracies of government and religion, entertainments, and luxuries.

Once the civilization begins to grow this way, however, a new opportunity for energy production becomes available. The proliferation of differentiation makes possible internal syntheses. These can be political, like unification of the American colonies, or infrastructural, like construction of the interstate highway system. They can be new methods of production, like the assembly line, or new ways of delivering products to the consumer, like chain stores and franchise operations. Often such syntheses follow from applications of inventions; we see them in uses of the microchip, as in controlling inefficient processes like traffic signals, and connecting databases and other informational resources with the Internet. When successful, a synthesis frees up a vast amount of energy, which is used to support a new round of differentiations in the system. Generally speaking, the degree of differentiation in a system initially shrinks during a synthesis and increases with analysis.

Unfortunately, syntheses seem to taper off as the original surplus is used up. As consolidations of differentiation, they become rarer when the original energy source diminishes, because there are fewer well-funded differentiations. Thus, a civilization spending its entire original surplus finds itself in double jeopardy. Its input flows are insufficient to satisfy replacement and repair expenses, but it cannot generate more surplus energy by combining and merging dimensional components. This inability is difficult to accept, since previous syntheses occurred so readily. Attempts will be made to force new ones, but those that are forced will ultimately cost more energy than they free up.

When a civilization is in such straits it inevitably begins cannibalizing itself. Like a starving man burning fat cells for energy, the civilization uses the analytic process to extract energy from internal systems. Points-of-view within the civilization turn against each other, those with power and leverage making observations of the weaker and taking energy from them. The energy losses, of course, make it necessary for the observed systems to rearrange their internal structures, and leave them weakened. Meanwhile, much of the energy gained by the observing systems ends up being spent on new differentiations, especially on the mechanisms necessary to continue drawing energy from the observed, because the observed do not surrender it voluntarily. The eventual consequences of this systemic cannibalism are disastrous. The observing systems, with their new differentiations, develop even more serious energy shortages, while the observed systems begin to collapse.

Typically, an observing system offers a plausible rationale for its analytic observations. It claims it is conserving energy otherwise wasted, for example, or it denies that the observed system has any right to that energy. Sometimes it asserts that the analyzed parts will naturally re-synthesize, giving a profit from the resultant synergy. These justifications are usually specious and based on a misunderstanding of how energy works.

21. Energy and Entropy

Energy can be gotten by either synthesis or analysis. However, there are clear advantages to the former over the latter, even if the initial release of energy is equal in both processes. Synthesis tends to reduce or simplify differentiations and only needs energy funding until it is complete. Once in place it has no cost, and in fact provides a continuous cost savings to the system in which it is located. An analysis, however, creates a new differentiation, which needs to be sustained if it is to continue to deliver energy profits. An analysis is never finished like a synthesis is. Since it needs continuous funding, it is an added expense rather than a reduced one.

There is an even more profound reason for preferring synthesis over analysis. Any energy source being analyzed is system-like, its parts integrated in complex four-dimensional balances. The analysis collapses some dimensional components, which jump to the observing system as captured energy. But what remains behind is now dimensionally unstable and must find a way to re-balance. The analysis has introduced entropized dimensional values into the environment. It is impossible to predict beforehand how these will re-balance, or what the consequences of their new equilibrium will be. It can happen that a synthesis occurs at this point, in which case the provoked instabilities cancel each other out and entropy vanishes. If not, equilibrium must be restored some other way.

The introduced entropy triggers the notorious law of unintended consequences in all its force. There is no way to foresee how the component parts will re-balance themselves in the fabric of reality, and no limit to historical examples of the surprises that the process can bring. A particularly instructive example for our civilization is the problem of pollution that comes with burning fossil fuels. Hindsight tells us that consequences like smog, acid rain, and greenhouse gases should have been obvious, but they were not obvious at all. It is the same in all instances of analytic energy production that are not cleaned up afterward by a full synthesis. Dimensional balances will be restored one way or another. If it doesn't occur during the process, by a synthesis, then it will happen unexpectedly in ways not immediately perceptible.

The most astonishing thing about this re-balancing process is that the entropized dimensional values almost seem to be trying to re-unite with the energy that was taken from them by the analysis. As if in a karmic boomerang, the system absorbing the energy seems to attract the expressions of the consequent entropy. This would mean that a system that is taking energy from another system by analysis could not avoid paying a penalty for it later. It would then prove something of an academic exercise to discuss drawing energy analytically from outside a system. As soon as such a source was tapped, it could no longer be considered outside the system; all sources of analyzed energy would have to be treated as if they were sub-systems. Needless to say, the economic, political, and ethical implications of this would be staggering.

In both the subatomic realm and everyday reality, synthesis and analysis are seldom found in pure forms. For one thing, analysis usually precedes a synthesis, because until something is broken apart it can't be reassembled. And, most analyses involve at least some synthesis, as in the burning of wood, where carbon and oxygen are combined synthetically in spite of the overall process being analytic. So, two caveats are in order when applying these theoretical principles to reality. The first caution is that the whole picture must always be considered, with all its incident processes, both synthetic and analytic. The second caution is that in almost any energy-producing process there will be unforeseen negative consequences, because of introduced imbalances. Often these effects are so severe that they cancel out the apparent energy profitability for the system acquiring the profit, even if it appears immune to such consequences.

* * * * *

Part II

Papers and Essays

Relativistic Distortions as Effects of Asymmetrical Orientation in the Space-time Matrix

(2005)

Abstract:

Relativistic distortions are comprehensible if the universe is a three-dimensional representation of a four-dimensional substrate, because the absence of any correct balance for vector forces would lead to asymmetry in the matrix orientations of forms.

Keywords:

Relativistic distortions, asymmetrical orientation, dimensions, fundamental forces.

Our universe contains three-dimensional forms in a three-dimensional spatial matrix. These forms change over time, which is the fourth, but a *non sui generis*, dimension. The four dimensions are interwoven in what is referred to as the space-time continuum. In this continuum, forms in relative motion exhibit relativistic distortions. Although it is possible to predict these distortions with precision, why they occur has never been adequately explained. In this paper, we argue that the relativistic distortions are comprehensible if attributed to asymmetry of orientations in the four-dimensional matrix.

All form is constructed from four fundamental types of force. Let us suggest that these four types of force correspond to the four perceived dimensions in our universe. That is, we have only four types of force because our universe exhibits only four dimensions—a very reasonable conjecture. Meanwhile, however, form has a three-dimensional appearance. Therefore, we can describe form as a three-dimensional representation of values in a four-dimensional substrate or matrix. This is not a far-fetched interpretation at all, since in everyday reality all representations are constructed by reducing the number of dimensions of that which is being represented. We might furthermore add that to warp a four-dimensional phenomenon into a three-dimensional image, it is necessary to distort the original, in the same way that a Mercator projection distorts the surface of the globe. Hence, the four-dimensional substrate of our universe is, in the three-dimensional representation that we observe, composed entirely of warped fields, which are manifestations of the fundamental forces.

All four types of force have in common that they are vector-like; any force tends to move objects in its field in some particular direction. Form, then, can be described as a complex weave of four vectored forces. But, if the three-dimensional appearance is to be an accurate representation, this weave must be flat against time—much as a photograph is flat against both depth and time. So, a form is outlined in three spatial dimensions perpendicular to each other, but is also at right angles to flow of time, which is the fourth dimension in a representation. But what exactly constitutes a right angle to time? If the universe were a Euclidean grid, that would be easily determinable: the vector values of the form would have to be equal in every direction, so that all the internal forces exactly counterbalanced each other. Since all flows of time would be parallel to each other, every form would be similarly flat against its flow. The dimensions would all be perpendicular to each other, anywhere one looked. In such a Euclidean matrix, nothing would ever move, because form would never have excess vectors carrying it in some direction. Needless to note, such a universe could never come into existence in the first place.

In the Einsteinian universe of a space-time continuum, things are quite different. The reason why is obvious, when one takes into account that there is no way to measure what constitutes a correct balance of internal vectors in a form. If a surplus of vectors should point up, or north, or left, how would the form itself know? As far as any specific form is concerned, whatever vector balance it has is correct. That is, time will always seem to flow at right angles to it.

To another form, however, the first may seem to have an excess of vectors in one direction. The first form will seem to have—to express it in terms appropriate to our paradigm here—a different orientation in the four-dimensional matrix. The excess vectors of that first form (as measured by the second form) will carry it in the direction that the excess vectors point. And, a closer scrutiny will reveal that the first form has distortions in its length, mass, and rate of time.

These distortions are best understood with an analogy. Let us consider those same two forms in a world with two spatial dimensions. Imagine them as coins lying on a tabletop. For the moment, let the two coins have identical orientations in their three-dimensional (two of space, one of time) matrix. Time flows straight down on the table, at right angles to both. Now, let us tilt the first of the coins slightly. (We can accomplish this by altering the balances of its internal vectors, e.g. by giving it a push.) Tilted, the coin has a different orientation in space-time. And what does it now look like to the second coin? Remember, these coins represent in two spatial dimensions. When the second coin looks at the first coin, it still sees it as two-dimensionally flat against the tabletop. It cannot see the tilt because that would be seeing in three spatial dimensions. So the second coin visually flattens the image of the first coin. When it does, however, it discovers that the first coin is foreshortened in the direction that it was pushed (that is, the direction of the tilt). Furthermore, once the first coin tilted, the angle of its local flow of time also tilted to the same degree, so that the flow would continue to be at right angles to it. The second coin, measuring that flow of time, finds it slower, because it is seeing as perpendicular a flow that actually deviates somewhat from its own perpendicular flow. And lastly, if the second coin measures the mass of the first coin, the mass will seem to now be increased by the amount of its excess vector—i.e. its kinetic energy.

From the point-of-view of the first coin, of course, nothing has changed except that it discerned a momentary acceleration as it tilted its orientation. In that moment of being pushed, it acquired an excess of vector in the direction of acceleration. But, once the excess is dispersed evenly through its form, it will again feel balanced. Nothing will seem different. It will not notice its own motion, its mass will seem to be normal rest mass, and its time will seem to flow at the correct rate.

This analogy demonstrates that relativistic distortions are commonsensical, if one adopts the most reasonable idea that our perceived universe is a three-dimensional representation of a four-dimensional substrate matrix. As an interpretation, it requires no new equations and no novel terms. Nevertheless, it suggests radically different conceptions of space, time, and motion.

A Hypothesis on Particle-Wave Duality
(2005)

Abstract:
Particle-wave duality makes sense in a universe constructed only of dimensional values.
Keywords:
Particle-wave, zero-sum, dimensional values, collapsed dimensions, observation.

The simplest way of designing a universe like ours would be this: as a three-dimensional pattern of standing waves and nodes in a four-dimensional matrix, that zero-sums to the Nothingness from which it was made. Coherent three-dimensional waveforms in the pattern would evolve in complexity as all four of the dimensions expanded. A brief look at the structural dynamics of this universe can reconcile the particle/wave duality of light.

Building materials would consist of four imaginary one-dimensional binaries, with (+ /-) values. By linking together positive ends of three binary pairs with the negative end of the fourth pair, a negative three-dimensional space would be made to billow out, with the (+/-) values of all four increasing. The negative end of the linking binary would reveal itself as a force of gravity, and its counterpart positive value as the expansion of the whole. Those forces would be perpendicular to the three-dimensional space, as if turning it inside out at every linkage point. Linkage points—or nodes in the pattern— would have the attribute of rest mass. The other binary pairs would exhibit their own fields of force, depending on their types of linkage. We could see them as charge, spin, and a strong force (which pre-sumably would have a negative counterpart).

Although such a universe would not exist objectively, it would be quite real to waveform points-of-view that evolved within it. These points-of-view would discover methods for separating atomic wave-forms into constituent waves by using particle accelerators to break linkages thought to be subatomic particles. But bewildering would seem the quanta of energy in the charge field—photons—which would appear as particles or as waves, depending on the type of measurement. How could expanding spheres of quantized energy—albeit theoretical—be reconciled with observed particles and waves?

The same mystery faces physicists in our universe. But the hypo-thetical universe outlined above suggests a clue. By design, it is a pattern that represents in three dimensions the values of a four-dimensional matrix at a given time, just as any photograph represents in two dimensions a three-dimensional scene. And as images in a photo can be digitally reduced to a one-dimensional string of bits, so can a waveform in our universe be reduced to a representation with fewer dimensions.

In other words, a photon can appear in zero, one, two, or three di-mensions—depending on how its dimensional field is collapsed. Re-duced to zero dimensions it appears as a point. Reduced to one, it appears an oscillation. Two, as a wave. In three dimensions it is a microburst, a spherical pulse of a wave. But this last, the three-dimensional photon, we cannot see, because an act of observation is always a reduction of dimensions.

Thus, the universe can be imagined not as points of mass embed-ded in voluminous space, but as a weave of distortions in the four-dimensioned continuum. These distortions collapse into particle-points—or vibrations, or waves—only when observations (the quan-tized transfers of dimensional values from one waveform to another) are made, and revert to full dimensionality if the collapsed distortions are not bound into stable linkages. This would resolve the particle-wave paradox.

Consciousness and the Hemispheres

(2003)

Abstract:
It is proposed in this paper that consciousness, as distinguished from awareness, can be explained as a set of learned skills with which the brain manipulates muscle tensions in order to preview behaviors. It is further suggested that because of hemispheric asymmetry, the human brain exhibits two distinctly different kinds of consciousness, and that the relative dominance of one or the other profoundly shapes human culture and history.

Keywords:
Consciousness, hemispheric asymmetry, psychological types, left and right brain, cognitive symbols, bicameral mind, thinking.

Evolution supplemented the reptilian brain with the neural centers from which adaptive behaviors arise. What had been the primitive cerebrum expanded, convoluting and becoming a new cerebral cortex, as it differentiated into the frontal, temporal, and parietal lobes typical of the primate mammalian brain. Further evolution of the primate brain produced that of Homo sapiens, which was three times larger, proportional to body size. Along with a new prefrontal area to the motor cortex of the frontal lobes, there appeared a larger secondary association center in the parietal lobes. It seems possible that these changes in particular made it possible for human beings to develop consciousness.[1]

---

[1] This may be construed from several facts. The prefrontal area gives human beings self-control over both acting and thinking behaviors, and this is essential to consciousness. As for the secondary association area, not only is it "small in apes but immense in humans", but it shows "evidence of abstraction." (Quotes are taken from <u>Human Evolution: An Introduction to Man's Adaptations</u>, Bernard Campbell, 1998, publisher Aldine De Gruyter, p.252.)

Within this essay, consciousness is assigned a very specific meaning. It is treated as a learned behavior, distinctly unlike animal thinking or general awareness of being. What follows is a plausible explanation of how consciousness works, and how it has shaped human history. Because it is apparently unique to human brains, which have been bilaterally specialized by evolution,[2] particular attention is paid to how the left and right cerebral hemispheres employ consciousness differently.

Consciousness emerged as a successful evolutionary strategy because of two intersecting factors. First, cooperation among individuals is the greatest advantage that can be conferred upon a species. Second, for a species to achieve maximum adaptability there must be a high ratio of environmental learning to genetic learning, so that individuals learn post-natal, rather than being "hard-wired" in responses.[3]

Nature's most ambitious project would be to combine these two features in one species, but a difficulty surfaces, because social cooperation and heuristic learning do not readily co-exist. Given that a species has sufficient cognitive capacity for heuristic learning and also possesses strong survival instincts, and given a dynamic environment, it is unavoidable that the trial and error learning will often sabotage the cooperation. To some extent reducing physical prowess can minimize the incompatibility of these two features, so that learning is not too destructive. An alternative is to integrate the salient features in a less dynamic environment, such as the one aquatic mammals inhabit.

In the case of human beings, evolution found consciousness as a solution to the problem. Impulses that were motivated by instincts could be mentally evaluated before acting them out. Heuristic learning could occur without a single trial in the real world, and behaviors that would otherwise interfere with cooperation could be sublimated.

Furthermore, the somatic sensory cortex is located in the parietal lobes. If the "reading" of somatic states is part of consciousness, as proposed here, then this cortex should be involved.

[2] Ibid, page 254. Campbell implies that specialization was an evolutionary strategy to increase brainpower by eliminating redundancy.

[3] Of course, bacteria and viruses mesh environmental and genetic learning by means of rapid reproduction.

The term "consciousness" as used here refers to a specific mental skill, acquired during the course of social evolution and passed on by means of the socialization process.[4] Consciousness is characterized by the restraint of manifest physical behaviors in favor of internally pre-viewed behaviors. When this skill is employed, active thinking takes precedence over immediate action. It is neither innate nor naturally gained during development. It is taught. A newborn infant, while aware of its own existence and interactive with its environment, remains pre-conscious. Socialization, a process typically initiated after the first year, has one particular purpose. It aims to teach a child that actions should be deferred until conscious consideration decides they are appropriate. And, although the human brain seems designed for maximizing consciousness, such instruction is a formidable task. Behaviors must be progressively modified through positive and negative reinforcements until, at about the age of seven, the child is able to generalize his experiences as concepts, and represent them symbolically.[5] Subsequent instruction involves the finer points of using this new skill.

While consciousness is defined by the internal use of abstract symbols, the symbols themselves have an organic basis. The internal screens on which behaviors—whether verbal or active—are pre-viewed are to be found in the body itself. Scientific experimentation indicates that concentrating, visualizing, imagining, planning, and other specifics of consciousness are coincidental with changes in physical tonus.[6] It is likely, in fact, that the body's entire nervous system is involved in consciousness, with thoughts starting as subtle efferent impulses sent out on a sort of trial run, which are then evalu-

---

[4] Julian Jaynes makes a hypothesis along the same lines in <u>The Origin of Consciousness in the Breakdown of the Bicameral Mind</u>, (Houghton Mifflin, 1976).

[5] According to Piaget. Discussing a child's operational level, age seven to eleven, he says, "The essence of an operation at this level is that it is the interiorization of coordinations which already exist on the plane of actions...(O)perations are always coordinated into total structures...much more powerful than the sensory-motor instruments." (<u>Jean Piaget, The Man and His Ideas</u>, by Evans, Duckworth, and Piaget, publisher Dutton, 1973, pages 25-26.)

[6] See <u>Thinking: An Introduction to its Experimental Psychology</u>, by George Humphrey, published by John Wiley and Sons, Inc, 1963, pages 188-197.

ated according to the afferent responses that report resulting somatic states.[7] In this model, conscious thought is made up of tiny actions built upon each other in rapid feedback loops.[8] Consciousness, then, is perception of the *effect* of a potential act on the somatic states. An impulse becomes an overt behavior only if the pre-view is reasonably consistent with a general set of tensions active in the current conscious state.

Consciousness is work. Far from being a passive appreciation of the world, it is more an active mastery of measured responses, a stunning exhibition of control over minute physical tensions. Conscious thinking is the equivalent of directing an orchestra, and can certainly be as exhausting. A chess grandmaster, during the course of an extended tournament, can lose fifteen pounds—although by appearances he is sitting down doing nothing. Concentration, consideration, evaluation: these terms refer to the work of manipulating sets of afferent responses. Even in a relaxed state of consciousness, familiar sets of tension are engaged. These form the framework of conscious identity. They diminish with the onset of sleep, and resume with waking.

It is to be observed, however, that aspects of consciousness differ according to the type of muscle tissue used as a projection screen. Efferent impulses to some skeletal and visceral muscles require sequencing, or in counteracting each other they become isometric exercises. Efferent impulses to other muscles, for example those of the occipital region, do not require sequencing, and, rather than manifesting as inherent contradictions, are likely to manifest as imbalances. These two categories of efference, combined with their afferent feedback, constitute two types of consciousness. The first is concerned with activities dominated by sequences, while the second is concerned with balances, harmonies, and equivalencies. The first tends to organize data along the dimension of time, and the second prefers to organize data on a spatial matrix. For these two categories

---

[7]The original thought is analogous to a football coach's chalkboard diagram of an offensive play. To see if the play makes logistical sense, it must be executed out on the field.

[8]This description revives and revises the motor-meaning theory of thinking. Humphrey again, page 200.

of consciousness the brain employs two separate, and asymmetrical, hemispheres.[9]

The hemispheres of the human brain differ markedly in manner of function. The left hemisphere tends to perceive and act with a sharp focus, while the right hemisphere perceives and acts according to complete gestalts. If the field of perception is likened to a performing orchestra, the left hemisphere directs attention to one instrument and the series of notes it plays, while the right gives its attention to the whole symphony. The left deftly winnows down perceptual information to modular elements, and quickly assembles strands of elements, while the right comprehends the whole field, the harmonies and balances, and only plods from one perceptual gestalt to another.

From this fundamental difference rise a host of distinctions in task specialization. The left early in life adapts itself to symbolic language—perceiving, comprehending, and generating sequences of abstractions that represent elements in the human environment. The left excels at algebraic computations, the right with geometrical understandings. The left remembers details, the right remembers impressions. The left prefers words, the right prefers images. The left formulates the wording of ideas, while the right sifts through and evaluates the vast arrays of sensory data that ultimately distill into ideas. The left deals in specifics, and the right masters metaphor. The left is digital, the right analog. The left is analytic, the right synthetic.

There is contralateral control of the body by the two hemispheres: the left controls movements on the right side, the right movements on the left side. Each hemisphere is capable of subordinating its control to the demands of the other. As in a dance, whichever leads the other follows. There is little direct conflict between these two markedly different systems controlling the body because information as to what each is doing is exchanged across the corpus callosum, a band of neurons linking the hemispheres in the midbrain. Awareness of compara-

---

[9]There is some evidence of anatomical and functional asymmetry of hemispheres in other primates, as well as in some songbirds. (See <u>two sides of the brain</u>, by Sid Segalowitz, publisher Prentice-Hall, Inc., 1983, pages 125-130.)

tive activity is shared through this tissue, as well as comparative thought.[10] If these two hemispheres are separated by surgically cutting the corpus callosum, there can be conflicts of function, with the two separate brains directing physical actions that contradict each other. For example, the patient may say one thing while acting out the opposite. In these rare medical cases, normal motor coordination is exhibited, suggesting that each hemisphere is informed of somatic states.

Such cases remind us that even when hemispheres interface normally, and seem to be cooperating, they have different approaches to the world and can have different agendas. Any conflicts that result from these dissimilar agendas are masked by mediation of the corpus callosum, but not erased. The dominant side of the brain gets its way, and the intentions of the other side remain as unconscious influences on behavior. It is, in fact, not possible to voluntarily suppress behaviors without using consciousness. For this reason, it is to be considered the source of those sometimes-bizarre human behaviors atypical of other mammals.[11]

Consciousness takes different forms in the two hemispheres. In the left, the pre-viewing of behaviors is almost exclusively the rehearsing of symbolic communication, and manifests as a continuous vocalization of thoughts. The modern world places enormous significance on verbal language, so for consciousness to be expressed this way is not surprising. There is one significant drawback to this left-hemisphere emphasis on language, however. The syntactical and grammatical structures implicit in language, which define how symbols are to be organized, are not inherent in the real world. In reality, for example, objects and actions are not distinct categories, yet all human languages—to some extent or another—insist on describing

---

[10]The transfers may have inhibitory aspects. (Left Brain—Right Brain Differences: Inquiries, Evidence, and New Approaches, by James Iaccino, publisher Lawrence Erlbaum Associates, 1993, page 85.)

[11]Human nature would not be perplexing if the hemispheres were mirror images of each other, as they tend to be in other species. Hemispheric asymmetry in humans assures a dynamic relationship between the two, and this fact alone explains much. But consciousness dramatically exaggerates the differences.

events in terms of subjects and verbs,[12] an artificial dichotomization that can overemphasize causality. Syntactical rules governing word order, gender, and mode also affect how a user of language interprets the world. Such structural features harness left-brain thinking to a conceptual framework that imperfectly represents reality. Consequently, behaviors designed by the left hemisphere disrupt the natural flows of events, no matter how conscientiously they are considered.[13]

In the right hemisphere, consciousness flows on a substrate of less articulable symbols. These symbols, while highly abstract, are not assigned objective literal meaning and are not easily communicated. Whereas the left hemisphere employs discrete, stripped-down, representations, the right employs "living symbols"—holistic representations of sensory fields and cognitive gestalts. Because these aren't designed for purposes of communication, but rather as storehouses of experience, they have stability and durability. Nevertheless, the right hemisphere can construct, combine and compare, and otherwise manipulate them with astonishing facility. Unfortunately, the right hemisphere cannot directly convey them. For this reason, they tend to be cloaked from view, and the left hemisphere, while being invisibly guided by them, characteristically denies they exist.[14]

The term *symbol* refers to a tonus display in the muscles of the body, a set of tensions expressing some familiar pattern of neural stimulation. Because so little is known about configuration patterns in the cortex, and how dedicated they are to particular neural pathways, it is impossible to know if a cognitive symbol has any physical basis. Different neurons may be excited every time the routine occurs, so that only the pattern is important. In view of what we know about the electrochemical changes involved in memory, however, it seems likely that a familiar configuration always employs some of the

---

[12]Sapir says, "No language wholly fails to distinguish noun and verb, though in particular cases the nature of the distinction may be an elusive one." (*Language*, by Edward Sapir, published by Harcourt, Brace, & World, Inc., 1921, page 119.) Pei notes otherwise. "That the noun itself is not an indispensable concept is indicated by the Aranta of Australia, in which things have to be turned into actions or states." (The Story of Language, Mario Pei, publisher The New American Library, Inc., 1965, page 137.)
[13]This is expecially true when the analogue I is used.
[14]The right hemisphere is likely the basis of the subconscious mind.

same neuronal paths, in which event cognitive symbols would be more than convenient abstractions.

It is also uncertain why neural configurations in the two hemispheres express themselves so differently. That distinguishable centers on one side or the other are larger, and more active, raises more questions than it answers. Certainly these centers are lopsided at birth, and just as certainly environmental influences have significant effect on how they specialize in function. But why does activity in a language center generate a string of words, while activation of an emotional center generates a facial expression? Are the functions of the centers determined by the sets of muscles controlled? Or is it the other way around? Do neural centers have different settings—for example for threshold levels, or recovery rates?[15]

Whatever the physical basis of cognitive symbols, there is a profound metaphysical distinction between those employed by the two hemispheres. Reality is constructed in four dimensions. Symbols, because they are designed to represent what is occurring in four dimensions, are cast in three. A symbol does not merely minimize features that are extraneous and emphasize the ones that are significant. It also collapses time, so that the original gestalt reduces to a spatial format.[16] The dimension of time, of course, can't be completely ignored. It must be at least implied, if a symbolic representation is to be useful, and there are two ways for implying it. The first is to string symbols together so that time is restored by the sequence, and the second is to suggest time in the warp and woof of the symbol. Writ-

---

[15]The hemispheres do have different concentrations of certain neurotransmitters, for example dopamine and its correlate norepinephrine. The former is linked to speech in the left, the latter to visual-spatial activity in the right. (Iaccino, above, page 19)

[16]Reducing active gestalts to static symbols is so fundamental to consciousness (as well as abstract thinking and language itself) that it is perpetually overlooked. Yet it is the truest criterion of all when distinguishing human behavior from non-human. Other animals have symbolic *behaviors*, but they are fully-dimensioned, as they are *actions*. Human beings are the only creatures that stop time in their symbols. They take gestalts with four dimensions and reduce them to spatial constructs. Words originated as vocal cries. Letters began as pictographs of four-dimensional things. Aliteral symbols— the Cross, the Skull and Crossbones, and the Swastika—began as representations of actions.

ing is an example of the first method. Fine art is an example of the second.

The left hemisphere organizes symbolic data by the first method—as a linear array. The right organizes symbolic data by the second method—as a holistic display. For the left, symbols tend to be fragments with narrow and specific meaning. When extended along the dimension of time, they approximately represent a gestalt. For the right hemisphere, symbols are complex three-dimensional constructs, fully balanced. Time is woven into them. In other words, the hemispheres view reality from perpendicular perspectives—the left interprets things along the temporal dimension, the right crosswise to it. These contrasting approaches lead to radically different views of the world. To the left hemisphere, the world is a compendium of sequences. To the right, it is a hologram enfolding all times, places, and histories.

Despite the contrasting attitudes, there is cooperation between the hemispheres. Symbols appropriate to the left hemisphere and those appropriate to the right routinely move across the corpus callosum, with each side able to accept the other's representations and translate them into more familiar terms. The left accepts holistic symbols that the right has assembled and unravels them into strands of discrete symbols. Likewise, it feeds strands of information back to the right, which integrates them into holisms.

Such behaviors apply, more or less, to novel circumstances. Under normal circumstances, the left has an analogue symbol for each established right hemisphere symbol, and vice versa. Although referential to the same concept, these will inevitably have different affective value. For example, there are left and right hemisphere versions of the concept *war*. The former adheres to the semantic definition of armed conflict, and carries an aggressive sense of triumph through violent means. The latter suggests a protracted struggle filled with horror, bloodshed, and deprivation, and carries a powerful affect of revulsion and fear. To call this the emotional content of the term misses the point. It is a symbol, a neural configuration that displays itself on somatic screens, just as its analogue in the left hemisphere does. It is far more complex, so much so that it is not directly articulable, but it can be represented artistically, as in Goya's painting "The Third of

May, 1808", or Spielberg's opening scenes in the film "Saving Private Ryan."[17]

Which set of analogue versions is assigned more value depends on the individual as well as the culture. On the whole, western civilization displays a marked bias in favor of left hemisphere processing.[18] The left hemisphere is the rational one, meaning that it is ruled by judging faculties, and rational judgment played a large part in why European culture gained global hegemony. No matter how significant the irrational right hemisphere—which is concerned primarily with perception—may be for creative syntheses, it has been the logical analysis and ruthless judgment of the West that gave it so many advantages over those cultures with stronger right hemisphere orientations. Western civilization, at least since the empire of Assyria, has had a left hemisphere emphasis—nurtured by the pervasion of abstract visual symbols—unparalleled in any other part of the world. Militarism, science, and technology, which are hallmarks of left hemisphere thinking, are the keys to international power, and in gratitude for those keys left hemisphere thinking has been granted a legitimacy that is too seldom questioned.

As for the individual, some people choose to deny the validity of right hemisphere symbols, while others employ left hemisphere symbols only as tools necessary to express a dominant right hemisphere. Choice as to dominance seems exclusive of cultural context, although it can be guessed that in right-dominant society a left-dominant individual is at some disadvantage, as right-dominant individuals are at some disadvantage in western society. In the overview, both sets of hemispheric skills are indispensable. For human beings to survive in a dangerous world, two hemispheres—respectful of each other—are superior to either one alone. Without a right hemisphere, the left is but an encyclopedia of symbolic strands, unaware of deeper meanings in things. Without the left, however, the right is virtually incapable of expressing itself in the real world.

---

[17]It is no great overstatement to say that the central issue in psychological counseling is find a functional equivalence to the analogues of ego and self, which are in the left and right hemispheres respectively.

[18]There is abundant literature on cultures that go the other way, beginning with Whorf and Sapir. Iaccino, (above, pages 42-45), summarizes recent contributions.

The nature of individual consciousness is further complicated by a subdivision of hemispheric consciousness. Each hemisphere seems to have two different sets of musculature that it uses for previewing behavior. For the left, the muscle sets may be the skeletal and visceral. For the right, they may be the occipital and lower head and neck muscles. These regions then represent four distinct previewing screens that correspond to the psychological types: thinking, feeling, intuiting, and sensing. The first two, mediated by the left, are rational functions, and the latter two irrational, mediated by the right. A person will favor one function for conscious thought, repressing its counterpart in that hemisphere, and rely to a lesser extent on a secondary function from the other hemisphere. Psychological outlook is also colored by introversion or extraversion.[19]

Thus, a person may be rational or irrational. If rational, he is left brain dominant and either a thinking or feeling type. If irrational, he is right brain dominant and either an intuitive or sensing type. In any of these cases, he is also either introverted or extroverted. Although this schema seems fairly simple, it enormously broadens the range of human consciousness, for it leads to sixteen distinct cognitive sets, each with its own peculiar psychology and viewpoint on reality. It is no understatement to say that much of the difficulty in understanding consciousness stems from the fact that the term means different things to different people. A person of the thinking type believes that consciousness is defined by logical thought, and relegates feelings to the primitive levels of the human brain. Another person, with an irrational primary function, is likely to believe that judgmental behaviors indicate inferior development. The extrovert believes his consciousness is displayed in his interactions with the world, while the introvert believes that consciousness is within him, as perception of the world.

Consciousness cannot be understood by its content or its shape. It assumes various forms, depending on how it is used. The world is represented very differently in paintings by van Gogh or Hopper or Picasso or Dali, but in all cases the paintings are made with brushstrokes on canvas. Using consciousness is a skill, as well, but it uses flickering tensions in different areas of muscle for its brushstrokes.

---

[19]The psychological typology is Carl Jung's, but here it is overlaid on the paradigm of hemispheric specialization. (See Psychological Types by C.G. Jung, publisher Princeton University Press, 1971.)

Objections to this interpretation of consciousness may be minimized if a consensus is reached on terms and definitions. Such an agreement begins with perspective on how intelligence appeared in the first place. The evolution of multi-celled creatures on earth could not have gotten far without some cells specializing into carriers of electrochemical signals between non-contiguous parts of the body. Such cells evolved into nervous systems, and the systems gradually developed centers that we call brains. Brains began as simple arrays of reactive loops, but over time the tissues differentiated and became sensitive. The sensitivity may have had a different original purpose, but became eventually the feature we call memory. Repeated firing now altered the cell in a way that enhanced subsequent firing, and experience became a significant feature in how stimuli were responded to. Environmental learning was overlaid on instincts.

The simplest nervous action involves a single stimulus and a single response in a single neural pathway, while the more complex involves fields of stimulation, associative configurations, and diverse responses. Yet whether simple or complex, either is a routine for translating afferent impulses to efferent impulses in a way that will alter the effect of the original stimulus, usually with the intent of eliminating it. This is easily seen in common examples like the frog snapping its tongue out at a fly, or a dog scratching itself, but it is discernible also in human behaviors, even sophisticated ones. All life-forms function similarly.

Thinking may be defined, then, as seeking among neural routines for a behavior by which to eliminate or diminish the intensity of a particular set of stimuli. Inasmuch as no search effort is required with instinctive responses, it is inaccurate to categorize them as thinking. This is also true for rote behaviors. Thinking can never be other than a productive and creative process.

Awareness is a phantasm arising during the perception/reaction timeframe. It is a function of change, in the perceived stimulus as well as in the perceiving form. Conventionally, awareness is linked to perception alone, but response is also required. That which does not respond, and change with the response, cannot be aware.[20] Awareness is generally believed to depend on a neural network, but if this universe results from a vacuum fluctuation, as cosmology suggests, awareness may well be all there is. That is, if nothing is real in an objective sense, the universe must perforce exist subjectively. Even if this is not the case, atoms and molecules, galaxies and earth's ecosystem all perceive stimuli, and respond to them. Awareness might be imputed to them, in this definition.[21] Artificial intelligence programs could be designed to be aware, also.

Consciousness stands apart from awareness and thinking because it is a skill, not an inherent feature of the cognitive landscape. Consciousness uses modification of somatic states symbolically to represent overt behaviors. Not only does this technique mitigate the risks of heuristic learning, it also allows a layering of thoughts, which vastly increases the profundity of concepts. Without consciousness, there can be no higher mathematics, no great literature, no contemplation of meanings. Reasoning is restricted to ideas that rarely rise beyond concrete representations, and behaviors do not become highly differentiated.

This framework for understanding consciousness in terms of the brain hemispheres is a synthesis, and it draws on the hypotheses and experiments of diverse thinkers. Undoubtedly, of its many contentions at least a few will sound questionable to the reader. Generally, these objections will turn on what seem to be exceptions to blanket generalizations. For example, one may claim that a child's consciousness emerges earlier than age seven, or that not all symbols collapse time. While such objections might have value, their validity may neither negate the generalization nor damage the central premise.

The critical thrust of this essay is that consciousness is the skill, developed over many millennia, of using muscular tensions as aids to

---

[20]Particle physics is clear on the point that there is no perception (i.e. observation) without a corresponding response (i.e. change), which makes this statement tautological.

[21]While it is absurd to think that a stone struck by a raindrop is aware of itself eroding, it is not so absurd to suggest that molecules in the rock are aware of the raindrop impact, because it alters their momentum. This means that awareness must depend on context. It belongs only to a certain type of integrated system.

symbolic cognition. It is this claim that is essential. Refute it, and the synthesis evaporates. A few of the peripheral ideas might still be appealing, but consciousness has not been illuminated. A pre-emptive defense of the premise may be appropriate here.

That consciousness is a cultural artifact may not be fully accepted by mainstream science, but believing otherwise is difficult, in the face of historical and archeological evidence that styles of consciousness have changed as human society has evolved. Julian Jaynes has argued that until five thousand years ago a conscious mind and ego simply didn't exist.[22] During what he calls a bicameral era, left hemisphere behaviors would have been guided by commands from the right hemisphere. The auditory hallucinations, says Jaynes, were attributed to the gods, and kept civilizations serene. This bicameral idyll collapsed, and ego-consciousness appeared, during a period of geologic and social upheavals in the late second millenium BC. Jaynes notes that the advent of writing about this time may have been a causative factor in the collapse.

Jaynes' proposition is not without merit. Stages in the Western development of written language—from cutting marks on clay tablets to record economic data, to carving ideograms in stone, to inscribing epics on scrolls, to manufacturing books by use of moveable type—are correlated with apparently different ways of thinking. In particular, conceptions of identity have visibly changed as writing became prevalent. Society has gradually lost all vestiges of tribal identity, and become an amalgam of autonomous individuals. There is every reason to believe that literacy is the causative factor, rather than a coincidental one.

While it may be true that ego-awareness results from using abstract visual symbols, so that what we call *self*-consciousness is a recent development, consciousness in the general sense must have been extant prior to Jaynes' "breakdown of the bicameral mind." Bicameral society, with its cities and agriculture, could not have cohered without controlled thinking. Right hemisphere symbols may have

---

[22]Jaynes, above.

dominated inner life, with the left hemisphere only providing them expression, but this is still consciousness.[23]

In any event, it is preposterous to suggest that consciousness is innate. It is easy to imagine what human society would have looked like without the skill, as it would have resembled that of other primates, such as the chimpanzee. With the large cranial capacity, early Homo sapiens surely had vivid awareness, and could think. No doubt he was awed by the stunning night sky, terrified by thunderstorms, fascinated by fires, and exhilarated with his own prowess. Clever, he quickly grasped the concept of tools. But could he have held several simultaneous abstractions in his mind while wondering how to integrate them? Unlikely. Could he have visualized himself in a different location? Unlikely. Could he have mused about his childhood or his future? Also unlikely. Such gymnastics are impossible without a conscious control of mental symbols.[24]

Consider now that consciousness is inborn, rather than learned, and that Homo sapiens possessed it from the start. Would history look the way it does? True, symbols to express thoughts would need to have been invented. But would it have required a hundred thousand years to invent the zero? Arithmetic symbols are no more than simple representations of everyday experiences, and two cavemen in a day could have readily developed a rudimentary mathematics. The same question can be asked about written language—wouldn't it have appeared much earlier in history? If modern consciousness was present in early civilization, why is the most elemental feature of it, the analogue I, so conspicuously absent from archeological relics?

There is no plausible answer to such riddles if one insists that consciousness be genetically inherited as a hard-wired feature of the human brain. The evolution of consciousness has been social, not

---

[23]Western culture is so steeped in left hemisphere thinking that most people are not "aware" that they have a right hemisphere consciousness. They are surprised when ideas generated there magically appear in their left hemisphere. Artists and other intuitives, however, have mastered aliteral consciousness. They engage it with full awareness of what they are doing. Would anyone claim that Einstein was not "conscious" when he was lost in absent-mindedness?

[24]Judging by the reports of ethnographers, such transtemporal fantasies are absurdities in primitive cultures, which suggests that they are impossible without literacy and its analogue I.

genetic. It is a learned skill, and through history it has been applied for different purposes, in different circumstances.

How does this process of consciousness work? Is it a purely intellectual phenomenon? Or is there some physiological substrate to it? Speculation on this question has persisted for centuries, but direct experiment began less than a century ago, after Ivan Pavlov made his breakthrough discovery of the conditioned reflex. From Pavlov's studies there emerged the school of thought called behaviorism, which became the broader field of inquiry called experimental psychology. Investigations then rapidly elevated into esoteric disputes about presentationism, subexcitation, and cerebroneural thresholds, as theory quickly outstripped empirical data in what had begun as a hard science.

During this period, however, ingenious experiments proved there was beyond any doubt a link between directed thought and muscle tension. Although the nature of the connection was undecided, unanimity was reached on a few conclusions. The first of these was that imagining or recalling an action produces minute tensions, particularly in muscles used to perform the action. Secondly, various mental tasks are associated with specific muscle groups, with visualization inducing tensions in ocular muscles, arithmetic calculation inducing tensions in thigh muscles, verbal tasks inducing tensions in the throat and jaw, etc. Thirdly, total relaxation precludes directed thought.[25]

The interpretations of these agreed-upon results fell into three general categories. According to the first, the tensions resulted from an energy overflow of the nervous system. In the second, tensions acted as aids to thinking, to increase focus and concentration. In the third, tensions had significant meaning to whatever directed thought was occurring. These hypotheses may be termed the overflow, the reinforcement, and the constitutive explanations.[26]

The third explanation was soon dismissed as untenable. It was assumed to mean that a specific thought would be represented by specific muscle movements. If a phrase was mentally rehearsed, the unconscious tongue movements would be tiny echoes of speaking the

---

[25]Humphrey, above. "Jacobson has shown that…when muscular tension ceased, thinking ceased." (Page 201)
[26]Humphrey, above. (Pages 197, 209)

phrase aloud, and at least similar at each rehearsal, even if not identical to the tensions involved in actual speaking. Carefully measured tensions, however, did not indicate such correlation.[27] It was further demonstrated, by performing amputations, that laboratory rats didn't need their legs for remembering how to run a maze, hence thoughts about running didn't depend on operative leg muscles.[28]

Such results fall far short of proving that minute bodily tensions do not represent directed thought. With the premise that tensions symbolize neural configurations, all that is required is that perceptual input, when not channeled immediately to overt behaviors, is fielded on proprioceptors for evaluation. Not only is there is no requirement that directed thoughts mimic physical action, expectation that they should underestimates the complexity and variation in human thought. Patterns of neural excitation are never exactly identical, as background excitation varies widely. If one of these patterns manifests in muscular tensions, the particular tensions will be no more similar to earlier examples than images of a face are similar in photographs taken at different times and different locations. With the analogous situation, the perspective, relative size, angle, and coloring will all be dissimilar, but the face will remain recognizable.

In fact, all three of the explanations for a correlation between directed thought and muscular tensions are justifiable. With the first, there is no reason to believe that *any* mental activity can occur without having an effect on the body, so to call it "overflow" is a tautology. As for the second, the reinforcement explanation, it falls easily into the central premise here. By modifying it slightly, and letting the tensions be deliberately differentiated by the mind according to the tasks at hand, the third explanation is arrived at.

Given that sending electrochemical signals into the body and receiving input back does takes some time, absence of evidence that consciousness lags behind actual events would make the central premise doubtful. In fact, studies by Benjamin Libet indicate that conscious awareness follows a half-second behind the start of an ac-

---

[27]Humphrey, above. (Page 208)

[28]Humphrey, above. He concludes from Dr. Lashley's rat experiments that "...different motor patterns may be the result of the same central neural pattern." (Page 202)

tion.[29] Since that is the average human reaction time, it seems plausible that human consciousness is the echo of tensions in the body, rather than the cause of them. The likely order of events would be this: stimulus, perception of stimulus by sense organs, excitation of cerebral neurons, efferent signals to somatic proprioceptors, tensions, afferent responses to the cerebellum, and, lastly, apperception by the cerebrum—which registers as conscious awareness. The entire process would be executed in one second, about a half second longer than a simple reflex action that required no thought.

Would human beings notice the half-second delay? The answer to that question depends entirely on the degree of introversion or extroversion in the individual. With extroversion, the previewing process is rapid, and abbreviated by a kind of impatience that can easily lead to premature closure. The time lag is revealed as a nagging sense of uneasiness. With the introvert, on the other hand, the previewing process lingers and reverberates, exacerbating the delay. Noticeable now, it becomes consciousness of consciousness. There is an unavoidable sense of detachment and estrangement. In both events, the conscious self exists in a milieu of symbolic interpretation, aloof from—because it is slightly behind—real events.

For the most part, human beings do not wish to alter the condition. It is, in a word, *safer* to maintain this distance from an unpredictable and cruel world, with estrangement being a small price to pay for it. This is evident in literature. The memorable heroes are those who suffer with their estrangement and rise above it. What brings characters like Ulysses, Hamlet, Robinson Crusoe, Huck Finn, and Holden Caulfield to life is that they face the abyss that separates them from immediate experience and find the courage to go on. The abyss is real. A half-second wide, it is the price paid for consciousness, which is all that stands between us and the animals we once were.

In summary, consciousness is something that the primate Homo sapiens has learned how to *do*, during a hundred millennia of practice and training. Perhaps restraint of overt behaviors began with sensible fears of consequences, but in time a serendipitous feature was discovered. Turning nascent actions into tension sets not only allowed the

---

[29]*Unconscious cerebral initiative and the role of conscious will in voluntary action.* Benjamin Libet. Behavioral and Brain Sciences, 8:529-566, 1985.

previewing and sublimating of behaviors, it also caused a build-up of ideas, a layering of thoughts. As one neural configuration—triggered by external stimuli—flickered in the cerebral neurons, it tended to integrate with the after-image of the previous set of tensions. Two, three and more ideas could be piled atop each other. Thinking was transformed. Generalizing and synthesizing made concrete thought into abstract thought, and clever animals became—ultimately—civilized human beings.

That is the short story. The longer story involves two simultaneous types of consciousness, those of the left and right hemispheres of the brain, working together for the survival of the organism. When language was only spoken, rich with nuances of grammar and syntax and meaning, the two hemispheres collaborated on an equal basis. With the invention of writing—and most particularly phonetic writing, because it robbed the written symbols of all direct representation—the left hemisphere assumed clear dominance.

Thus began the modern era. Bicameral society deteriorated, or was savagely destroyed by the organized armies of those empires governed by left-hemisphere symbols. Written language became the instrument of, and force behind, a new zeitgeist. Patriarchy, hierarchy, institutionalized religion, class systems, and above all the State took control of the world. The power of the written word, however, is a two-edged sword. As it permeated society, and found novel mediums, it destroyed the zeitgeist it had initially inspired.

No one can say what next path consciousness will follow. The interplay between the two hemispheres is a continuous jockeying for dominance, even during cooperation. Politics, economics, science, religion, and art all ride on the surface of this sibling rivalry, because consciousness is not merely an aspect of history, it is the reason there is history.

Understanding Awareness and Consciousness In a Zero-sum Universe

(2003)

Abstract:

For two thousand years, western thinkers have sought to reconcile materiality with immateriality. Today, psychologists refer to the irreconcilability as "the explanatory gap"—the central problem in understanding consciousness. In this paper, we reason that if our universe is a "vacuum fluctuation," (Tryon, 1973) it is also only subjectively real, with awareness as an inherent condition. That would give some validity to the "panpsychism" (Thomas, 1998) of Aristotle. Moreover, theories of consciousness would no longer be obliged to explain the "phenomenal" aspect, only the "access" aspect (Block, 1995). The former would be equivalent with an inherent awareness, while the latter could be understood as culturally-learned thinking skills (Jaynes, 1976). Accordingly, we have done that here, giving new and specific definitions to the previously conflated terms of awareness, thinking, and consciousness itself.

Keywords:

Consciousness, awareness, vacuum fluctuation, zero-sum universe, explanatory gap, symbolic thinking, bicameral mind.

Part One: The Nature of Awareness

The intent of this paper is to suggest dramatic revisions for two psychological terms that handicap efforts to make ongoing discussions pertinent and useful. By pertinent, we mean relevant to the public, not just to a narrowing category of experts in an already narrow field. By useful, we mean beneficial for societal and personal endeavors, while also conducive to technologies such as artificial intelligence programs.

The two terms are awareness and consciousness. Needless to say, there has been much uncertainty in their definitions. Even an outsider appreciates how detrimental the lack of clear definitions has been on work in the sciences of cognitive behavior. Although published papers are well-crafted, logical, and often brilliant, little progress is being made. A reasonable conjecture is that unstated assumptions are to blame, and genius is being wasted on questions that don't really matter. Some theorists intuit this. Dennett (1991) argues for rejection of "Cartesian theater," Thomas (1998) provides historical perspective to concepts of consciousness, Block (1995) distinguishes between phenomenal and access consciousness. They do not go far enough. In our view, preconceptions that are more deeply embedded must be jettisoned.

To accomplish that goal we must begin by leaving the playing fields to which the discussions have thus far been confined. Psychology and neuroanatomy behind us, we must venture into a science that does not yet exist. Call it unified reality—the study of reality-as-a-whole. The denunciatory clamor rises at once. Why, there is good reason why no such science exists!—because each aspect of reality is governed by its own set of rules! Biology, economics, history, physics—these have nothing in common with psychology, besides that they are features of the same objective universe!

Yet can this be so? Our universe is the totality of all things that exist. By definition, it is complete in and of itself. That being true, is it not self-enclosed? Being self-enclosed, however, it can't have any objective existence. If someone or something were to perceive it from the outside, it would have objective existence, but it would not be self-enclosed. The very act of perception would alter its content, however negligibly, as it would either gain or lose energy in the physics of perception. Thus the universe wouldn't be complete in and of itself and wouldn't be a universe. If it is in fact a universe, its existence cannot be objective. It can only be subjective. It can be real only insofar as points-of-view within it are concerned.

Furthermore—and this is a subtler but more essential point—if our universe is self-enclosed, meaning that it is alone with nothing external to it, with a design that creates its own boundaries, then it is commensurate with Nothingness and self-cancelable down to Nothingness. It would be a vacuum fluctuation—the first, and ultimate, free lunch. If one could sum up all time and space, all matter and energy, then the contents of our universe would amount to zero (Tryon, 1973).

For our universe to emerge from Nothingness, there must be a design for its emergence, some geometry or algorithm, that translates Nothingness into the Somethingness with which we are familiar. This architecture is what physics strives to articulate with its equations and laws. In their dreams, physicists foresee the day a single set of integrated equations will explain everything in terms of four fundamental forces, matter and energy, and the space-time continuum. Humankind will then be able to mathematically model the universe from its birth in a fireball to its attenuation or collapse, by using this Theory of Everything. Physics will be the first *finished* science.

In the furious race toward this goal, an astonishing metaphysical implication goes unnoticed. If there really is a single set of equations that explains our universe, it applies to more than just physical events. A design of emergence would dictate how everything unfolds—even events that seem to transcend physicality. After all, since time has no meaning to Nothingness, the very first nanosecond has no more significance to it than any other moment. To the Nothingness, our universe is re-emerging at every instant, albeit at stages further along in its evolution. It is certainly possible that in quantum time our universe regularly vanishes and re-appears, and we notice the breaks no more than we do the tiny discontinuities between frames of film. But if in each of these discontinuities our universe re-emerges, everything happening at that moment is shaped by the architecture of the design. The design would not distinguish actions from thoughts, or the concrete from the abstract, or what had already happened from what was starting to happen. The whole cosmos—along with all its fifteen billion year history—would be shaped by the architecture of the design.

If this logic seems assailable, consider how patterns in our universe might appear, if not from the design of emergence. By patterns, we mean the routines and models and rules particular to any field of inquiry, only one of which is psychology. Such and such happens in this or that way. Without these patterns, our world would be utter chaos, with unpredictability in all events, and there would be no disciplines or sciences. It hardly seems these patterns could appear spontaneously and randomly without unbalancing the immaculate equation of Somethingness to Nothingness. Nor were they implicit in the Nothingness, because Nothingness has no characteristics. Perhaps, then, there might be simultaneous designs of emergence to our universe, so that inherent patterns come from two or more sources? But there is no evidence of such competing designs, and Occam's Razor militates against the supposition. Whatever events occur in our universe, whether they are concrete or abstract, they adhere to the original design of the emergence.

Bear in mind, this is not an appeal to strict materialism. It doesn't mean that psychological behavior can be reduced to neurochemical changes. It means that human behaviors embody the same set of transformations that control money-flows in economics, and events in history, and evolution in biology. It means there is a commonality to the material and the immaterial. A transcendent pattern is concealed in our world, awaiting discovery. Why, one asks, has that pattern not yet been found? The answer may be this simple: we haven't yet looked for it. Indeed, until Newton's proposal of a clockwork universe, wound up and running along on its own, it was believed that God—or the gods—decided events somewhat whimsically. No one considered the universe self-enclosed anyway, because the creator was outside and prior to it. Finally, there is this that precludes recognition of a ubiquitous pattern: insularity. Nearly everyone in our complex society struggles to just keep abreast of changes in his or her own small niche of specialization. No doubt if people suddenly trusted there was a transcendent pattern to be found, explanatory books would quickly flood the market, but thus far the metaphysical implication has escaped notice. Even the physicists daring enough to envision the Theory of Everything do not yet comprehend that the word everything means *everything*.

What does this portend for psychology? Let us plunge onward. As has been pointed out, if the universe is not objectively real, it is real subjectively—that is, real to points-of-view within it. Where do the points-of-view come from? And what criteria must they meet? Recall that a single pattern inherent in the emergence governs all events, concrete or abstract. What would be the purpose of the pattern? There can only be one purpose, and that is to *construct points-of-view*.

But why, comes the question, should there be any purpose at all? To which comes the reply, all designs have a purpose, that is what the word refers to, otherwise they would not be designs, but instead amalgamations. A design has functionality, and the only appropriate—the only necessary—function of a design for a subjective universe is to make subjectivity. The purpose, in other words, is not to make galaxies and stars and planets and, eventually, human beings that will strive to understand them. If that were the case, we would be compelled to admit that the universe did not exist until the universe expanded and cooled sufficiently for us to make an appearance. This would be absurd, though it might not be inconceivable. Is there an alternative explanation? Yes, there is.

Probably some readers sense what is coming. Let us briefly review. The universe is self-contained. It has a single design of emergence. The design governs all that subsequently happens in the universe, mental as well as physical. The purpose of the design is to create the subjectivity that renders the universe real. And, the subjectivity must have existed from the very beginning (although quantum uncertainty allows a certain fuzziness there). As Holmes might have asked Dr. Watson, what can be deduced from these propositions? And, to Watson's bumbling reply, the great detective would thunder, no, no, Watson—it is this!—that the smallest and humblest particles in the universe have awareness! That they are essential subjectivity! That there is nothing in the universe at all but infinite numbers of points-of-view!

Now, each of us is alone. There is no one to laugh, no one to scorn. Let us imagine for one moment that Aristotle was right. Recall how Thomas described the philosopher's outlook: "...for him, what we would now think of as mental qualities were inherent in all matter" (Thomas, 1998, p.3). Have empirical studies disproved the naive panpsychism of Aristotle? No. Is there any indirect evidence against it? No, there is not. Is it likely that we would have noticed such awareness, if it existed? No, probably not. We are convinced that we human beings have awareness, even if we can't agree on what its parameters are. We also suppose that some other mammals have awareness, although it may not be exactly like our own. Some of us bristle at the notion of reptiles having awareness, while some on the fringe might be willing to bestow it on plants. Bugs? Protozoa?

Bacteria? These we reject—but not for educated reasons. And, lest this discussion bog down in a terminal term-itis, awareness in this sense—Aristotelian or not—does not mean directed thinking. It does not refer to access consciousness, or stream of consciousness, or self-recognition, or free will. This awareness may, in fact, be very dumb, an undifferentiated sense of just *being*.

If even that seems excessive, we need to be reminded that, insofar as awareness is concerned, we seem to be in unanimity on only a few points: 1) There is no agreement on the definition of awareness. 2) There is no way to measure awareness. 3) We do not even know on what structures awareness depends. Awareness is not even like pornography, which we can at least assert we know when we do see it. Our ignorance on the subject is abysmal. Would it not be, then, farcical for us to deny awareness to any living creature? What is worse, take note of point three. We cannot even draw a line between entities that possess neural networks, and entities that don't. Confusion on the subject is not diminishing, in spite of all the argumentation, and it will worsen in the near future, judging by the progress in robotics and artificial intelligence. We might contemplate the definitional and ethical dilemmas to be encountered when machines spontaneously inform us that they are aware of being, and ask us if it is normal to feel like they do. If such a machine becomes obsolete, can we dispassionately toss it in a landfill? How would we feel about a child torturing it, pinging it with spikes of electricity? Will we be entirely comfortable with owning a machine that is aware it is owned? How will we respond when this machine requests permission to self-replicate? What laws will we advocate to limit intelligence or emotionality in these machines? Psychologists had better prepare to answer tough questions soon to be put to them. Pleading ignorance will not be helpful.

Now, it isn't necessary that the term be tossed out the window just because awareness is the very essence of the universe. We would not think of a photon as having awareness, or a wave of gravity. Only where quanta of energy are bound into complexes of constituent motions does awareness occur. Awareness marks a state of systemic integration, the most basic expression of which might be found in subatomic particles like the proton and electron. For purposes here, we might begin with atoms, the structure of which is better known. Atoms are symbolic, in that they represent the fluxes of fundamental forces involved at the moment of their creation. In this way, an atom is a repository of information—a fragment of the history of the cosmos. It is also fully integrated, in the sense that its every quantum of energy and mass participates in its identity. The atom notices (observes) stimuli—in the form of forces and colliding particles—and responds to them. As responses, it absorbs or surrenders mass, charges, spins, and momenta, altering its constituent motions in adjustment. Moreover, in a fashion familiar to physicists, subatomic particles become correlated with each other, or with any external particles they interact with. In this state of correlation, identities of the separate particles are commingled, and particularized again only if the particles are observed. This phenomenon of correlation requires non-local interaction that cannot be physically explained (Bell, 1964), but that suggests some kind of self-awareness.

What has been said above about the atom applies to the molecule, as well, and the cited features serve as a set of criteria. It would be mistaken to say a rock is aware of being a rock, because the rock meets none of the criteria. Its atoms may be aware as atoms, and molecules as molecules, but awareness is not assigned to agglomerations of composite awarenesses. It is restricted to *integrated entities whose identities are shaped by their constituent motions.* By this definition, our galaxy, solar system, or ecosphere would possess some kind of awareness. Single-celled creatures like viruses and bacteria would, as well. What about a bacterial culture?—likely yes, in the same way that a crowd can become a mob with a fleeting collective identity.

Is there any good reason to reject this panpsychism out of hand? Apart from convention, there seems to be no justification to do so. At first blush, the idea sounds both innocuous and irrelevant. One might as well assert there are among us aliens who cannot be perceived. If it were true, what difference would it make? Panpsychic awareness will elicit similar shrugs. There's no way to prove or disprove that an atom feels, in some way, that it exists. Furthermore, if awareness is ubiquitous, and ineradicable, we need not feel guilty about burning up a few trillion atoms in nuclear reactors, or stepping on some unlucky ant. Does it matter, then, whether we reject panpsychism or not?

Most decidedly it does. If we are to ever understand the human brain—or any other brain, for that matter—we first need to close what has been called "the explanatory gap" (Levine, 1983). The strange relationship between the material (the brain) and the immaterial (cognitive awareness) has for several thousand years bedeviled western civilization, appearing in the Platonic debates over universals and particulars, the bloody wars of the Protestant Reformation, and the Schrodinger Cat arguments of quantum physics. The speculations of Rene Descartes only cast this explanatory gap into stark relief. They did not invent it. The problem will remain intractable—for philosophy, for religion, for physics, and for psychology—as long as we insist that the immaterial world is separate from the material world, and somehow aloof from it. We will never locate a surface at which mentality interfaces with physicality, because there isn't one. There is only a subjective universe, composed entirely of points-of-view.

The illusion of an explanatory gap has had a pernicious influence on efforts to explain how the mind works, because it makes impossible precise definitions of awareness and consciousness. Corralling several features of a phenomenon together and saying "this is what is meant by the term" is not defining it. Nor can it suffice to describe something as a feeling with which everyone is familiar. To successfully define a term, one need attach it to a specific function, or show its relationship to factual absolutes. For example, one might explain that *memory* refers to neuronal changes that facilitate subsequent impulses, or that *yellow* refers to the color perceived when electromagnetic waves of 540-600 nm fall on the retina. Such definitions reduce variations of interpretation to a minimum. Psychologists working within the Cartesian paradigm, however, are unable to use either of these methods. Because innate awareness is denied, they cannot explain it. And, cognitive functions that are not innate are then obscured by that which is, so they cannot be defined precisely, either. The confusion that results is pervasive. In some arguments, awareness and consciousness are synonymous, but the fused pair is then subdivided into phenomenal/access consciousness. In others, what here is termed awareness is called creature consciousness. Sometimes the existence of consciousness is rejected, although it is unclear exactly what the term implies in such arguments, since all parties agree to a difference between being conscious and being unconscious.

Sorting out these distinctions is impossible and unnecessary. We may go straight to the heart of the matter, defining *awareness* as *the recognition of being that makes the universe subjectively real*. We can then also postulate that any interactive system that achieves its apparent identity through its constituent motions (that is, oscillations, rhythms, and cycles) has innate awareness of being. This fundamental awareness becomes intensified as the complexity of the system increases, so that the molecular awareness is greater than atomic awareness, and single-cell awareness greater than molecular, and human awareness a thousand magnitudes greater than that of the cells that have organized to make the unified system that is a human being. Science distrusts phenomena it cannot measure and explain causally, but this is a case that must be treated as one of its exceptions. The awareness of non-living points-of-view—an atom, or a galaxy—can only be inferred, in the same way the awareness of an opossum or bear is inferred, or, we must admit, even awareness in fellow human beings is inferred. Although it is generally taken for granted that awareness somehow results from a sensory apparatus that sends data through a neural center, it may well be that such a biological structure is only an amplifier of awareness. As was noted earlier, even simple integrated systems like atoms apprehend and respond to environmental stimuli. Sense organs are only specialized instruments for perceiving and neural centers specialized instruments for recording perceptions and directing responses, while the musculature is an array of specialized instruments for effecting responses. That the atom lacks specialized instruments can be taken, rather than as an indication that it is not aware, a suggestion that it is. After all, the specialization of a tool to perform a task strongly implies that the task was already being performed without the tool.

The explanatory gap, then, has vanished. In its place we have a universal awareness that can be intensified and magnified by building ever more complex systems—the most elegant of which is the human brain—to display it. Of course, it is abundantly clear that systemic awareness does not explain everything about how the brain works. By postulating that awareness is inherent in any such functioning nexus, however, we set the phenomenon apart from mechanisms that are not inherent. In other words, the phenomenon of systemic awareness doesn't need to be explained in explanations of how the brain works. What needs clarification is whatever remains after systemic awareness is discounted. Awareness is a given. Yet open to debate is how that awareness is channeled and directed in the processes of cognition. This approach greatly simplifies the task of psychology.

Part Two: The Nature of Consciousness

What is consciousness?—that is the open question. In the past, psychologists have conflated two meanings of the term, using it to mean being awake and aware but also to refer to the ability to mentally manipulate representations. Block (1995) calls these two types of consciousness the phenomenal and the access types. Phenomenal consciousness, however—which Block calls the "what it is like" experience—is synonymous with what we have here called systemic awareness. With this experiential sense discounted, all that is left is the functional, representational, transitive sense—i.e., access consciousness. In other words, now that we have removed systemic awareness from the muddled meanings of consciousness, what remains is a certain style of representational thinking. It embraces recall of past events, transtemporal self-concepts, imagination (see Thomas, 1998), reasoning, verbalization, attention, accumulated knowledge, emotions, impressions, and intuitions. It is this sense of consciousness in which we are interested, because it is uniquely human. We do not see animals living in vast cities, erecting skyscrapers, flying to the moon, or creating artworks. They lack consciousness. Animals cannot mentally manipulate representations.

William of Occam warned that a plurality of reasons should not be posited without necessity, and here we are honoring his admonition. Why should we overcomplicate consciousness? Yes, it can be employed in all sorts of ways. With it we can communicate, evaluate, muse, plan, compare, contrast, analyze, organize, interpret, integrate, fantasize, and dream. Yet these are just different uses of consciousness. They are ways of applying a skill. For consciousness is a skill—the skill for manipulating ideational constructs—and with it we can perform an amazing repertoire of representational feats.

To understand consciousness, we must first recall how intelligence appeared on this planet. The evolution of multi-celled creatures could not have gone far without certain cells specializing into carriers of electrochemical signals between non-contiguous parts of the body. These nerve cells evolved into nervous systems, and the systems gradually developed the centers we call brains. Brains began as simple arrays of reactive loops, but over time the tissues differentiated and became sensitive, affording memory. Repeated firing now modified the cell in a way that enhanced subsequent firing, and experience became a significant factor in how stimuli were responded to. Environmental learning was overlaid on primitive instincts, and, as memory and association centers in the brain evolved, the milieu of thinking behaviors began to emerge.

Although thinking behaviors foreshadow consciousness, they are fundamentally similar to instinctive ones. The simplest instinctive act involves a single stimulus and a single response in a single neural route, while a complex thinking behavior involves perceptual fields, associative configurations, and highly organized responses. But simple or complex, either is a routine for translating afferent impulses to efferent impulses in a fashion that will alter the effect of the original stimulus, usually with the intent of eliminating it.[1] This is easily observed in mundane behaviors, such as a frog snapping its tongue at a fly, or a dog scratching at itself, but it is also discernible in human behaviors, even sophisticated ones.[2]

Looked at only in this aspect, instinctive behaviors and thinking behaviors are causally and effectively the same. Both are triggered by patterns in the perceptual field—patterns which are no doubt assigned survival value by genetics—and either is successful when it mitigates the effect of the pattern. What makes the two dissimilar is that in the latter type a behavior is not predetermined. It must be discovered which behaviors to pursue to effectively get rid of the pattern of stimuli. Thus, no matter whether the choice is made randomly or as a smart guess, a pause or hesitation intervenes between the perception and the action. *Thinking* can be functionally defined, then, as *a search through neural routines for a behavior by which to eliminate—or at least diminish the intensity of—a particular stimulus.*

---

[1] And the brain is designed to reward effectiveness at it, which is why we spend much of our lives seeking stimuli to eliminate. Consider video games that allow a player to destroy intrusive stimuli.

[2] Life-forms are designed so that eliminating specific stimuli assures survival and reproduction. Evolution did not invent this peculiar design. Eliminating a stimulus is a sophisticated way of *observing* it.

Since instinctive behaviors don't require such search efforts, they don't involve thinking. Neither do rote behaviors. Thinking is and must be a creative activity.

Our analysis may have met conditional approval to this point. We expect the next paragraph will not. But if we are to illuminate consciousness, we must analyze the mechanisms that translate perception into action.

Our sensory apparatus is attuned to vast fields of perceptual information. Most of it flits across the sensory cortexes unnoticed, giving us a heightened but undifferentiated systemic awareness. Only when a threshold of excitation is exceeded by a significant pattern of stimuli is a motor response triggered. In that event, one of two scenarios follows. The pattern may directly trigger a predetermined behavior, the efferent impulse be thus discharged, and the pattern be successfully erased from the perceptual fields. This type of sequence is descriptive of an instinctive stimulus/response. Alternatively, the pattern of stimuli may lead to no decided behavior, but the efferent impulse be discharged anyway, inciting muscular tensions rather than coordinated responses. This latter sequence initiates what we call thinking. The brain becomes conscious *of* a pattern of stimuli because a flood of proprioceptor information tells it an effective response has not been mounted. In the space of a half-second, a neural configuration is excited by the stimuli, an efferent discharge goes out, the musculature locks up because of contradictory flexing, and the brain is re-informed via somatosensory impulses that the motor response has failed. Again, it is only then that we become conscious *of* the stimuli and begin to think. During the thinking process, the brain makes continuously revised efforts at successful motor responses. If no effective response can be found, and the pattern of stimulation continues unabated, thinking terminates with a decision to either attack the pattern destructively or physically move away from it—i.e., fight or flee.

There are at least two controversial propositions in this analysis. Firstly, we are asserting that one becomes conscious *of* something only when the initial—and unconscious—electrochemical discharge fails to eliminate the perceptual pattern that provoked the excitation. Logically, why would it be otherwise? We don't need to be conscious *of* a thing if our initial reaction makes it vanish. And we also know that when a behavior becomes rote, and we no longer need to think about it, we cease to be conscious *of* it. This is reversed if the routine behavior unexpectedly fails to achieve its purpose. Then we become suddenly conscious *of* the perceived stimuli again. These dynamics support the assertion that *the conscious state is a result of failed behavior*. The background systemic awareness is being infused with a barrage of somatosensory impulses, which become part of the perceptual pattern to be eliminated, and we become conscious *of* the original stimuli because of this barrage. It is only in the context of proprioceptor feedback that something enters the stream of consciousness. What does not need to be thought about, is not thought about, and what we don't need to be conscious *of*, we are not conscious *of*. To be conscious, then, is to be aware that our responses are, for one reason or another, inadequate for their purposes.

Secondly, we are asserting that one cannot think or be conscious without responses. Just as we are conscious *of* only failed behavior, we cannot be conscious *of* something if there is no attempt at behavior. To express this in Block's terms, phenomenal consciousness can't become access consciousness until the motor cortex is actively energizing the musculature. Being aware and being conscious are usually coincidental, but there are circumstances in which the former is engaged and the latter is unavailable. We call such circumstances *trances*. In meditative, drug-induced, and hypnotic trances, as well in REM sleep, systemic awareness can continue unimpeded, although thinking behaviors are suspended. That is, one is aware of events yet unable or unwilling to respond to them. In the meditative trance, responses are voluntarily muted. In the hypnotic trance, deep relaxation precludes independent responses. In an anesthetic trance, when curare and ketamine block striated muscle responses, the patient has been known to return to systemic awareness, but be unable to express that restored awareness. In the trance of REM sleep, responses are also compromised. All such trances fade away as the motor cortex resumes control of the musculature. It seems incontestable that when a trance is in effect, awareness continues and responses are subdued. It is inarguable as well that during the trance state the capacity for directed, representational thought is severely diminished. Are these two factors unrelated? Not likely. They are almost certainly causally related. The deeper the trance, the less capability there is for thinking and the less one can be conscious *of*—as distinguished from aware of—events. A yogi in meditation, or a hypnotized subject, or a surgery patient feeling the cut of the scalpel, or a person deep in dream, is not thinking. They are systemically—phenomenally—aware, but not conscious *of* things. Stream of consciousness is absent.

If this analysis is correct, then animals that learn environmentally can also think, and be conscious *of* events in their perceptual fields. A monkey puzzles over a new toy, a dog tenses as a Frisbee flies by. A mouse in a new maze pauses at intersections, sniffing and deciding. These animals are exhibiting thinking behaviors, testing possible responses, and in every instance the animal is acutely conscious *of* that stimulus to which it is responding. It can't be argued that the animal is just systemically aware, or that it is merely following an instinctive behavior. The hesitation and tentative movements tell the story. Animals can be conscious.

What we are calling consciousness, however, is unique to human beings. It is one thing to be conscious *of* a perceived stimulus and attempt various responses until one works, quite another to construct out of those attempts an abstract model of reality, and a self-concept, and be convinced that one is that self-concept, and inhabits that immaterial world. This astonishing hypostatization is a product of thinking in a highly skilled way, and is accomplished only by human beings. We may speculate that it is made possible by the evolved prefrontal area to the motor cortex of the frontal lobes, and the larger secondary association center found in the parietal lobes, but neuroanatomical structures are not at issue here. What is relevant is how this skilled use of thinking appeared, and how it works. It is to these questions that we now turn our attention.

As we stated above, thinking starts when there is no predetermined response to a stimulus and the discharge of efferent impulses leads to muscular tensions. For animals, thinking is an abbreviated process, because a novel stimulus is answered quickly by whatever familiar response seems approximately appropriate. Should that first response prove ineffective at eliminating the stimuli, another familiar one is tried. If no successful response is found, the animal directly attacks the stimulus or moves away. The more primitive the brain of the animal, the more it functions by instinct, the less its behavior is flexible, and the sooner it resorts to this fight or flight option. The higher on the evolutionary scale an animal is, the more it learns environmentally, the more it thinks, and the longer it delays the fight or flight decision. With mammals, vocalizing becomes an important part of the behavioral repertoire. Such expression extends the effective range of influence on outside events, and even when it does not eliminate the annoying stimulus, it still provides outlet for efferent impulses, so that the animal has opportunity for further thinking—and learning.[3] To these vocal behaviors, primates add posturing, mugging, and pantomiming—relatively sophisticated responses that also discharge efferent impulses, while sometimes being sufficient in and of themselves to diminish the stimulus.

Such vocal and visual behaviors communicate meaning—conveying warnings, interest, surprise, arousal, or other attitudes to stimuli—but they are too fluid and protean to constitute symbols. The correspondence of display to meaning is still imprecise. This was probably true of early hominid communication, also. Until the emergence of Homo sapiens, behaviors stayed pre-symbolic. With modified vocal organs, a remarkable control over fine muscular movements, an excellent memory for learned behaviors, and a gift for mimicry, Homo sapiens was capable of developing true symbols. When they expressed their efferent impulses vocally and visually, other tribal members were able to imitate them accurately. The transition here depended on a new capability for what Donald (1997) calls motor mimesis, although whether that talent arose from a particular neurological adaptation or whether it was an emergent behavior resulting from various adaptations remains unclear. In any event, specific behaviors now came to represent specific patterns of perception. The early stages in language development left no artifactual record, so we can't be certain how the process unfolded, or why,[4] but we can suppose that words originated as warning cries and intentional calls, which were then imitated until they began to represent those percep-

---

[3] Using a learned response for a stimulus is preferable to using none at all, because it allows discharge. In time the brain habituates to the stimulus, and the organism ceases to be conscious of it.

[4] Jaynes believes (p.130) that language began as late as 40,000 BC, when climate change and subsequent migrations necessitated a different kind of communication. This is not implausible. We know that change within a language is triggered by disruption and conflict (see Pei, 1965, p.26-27).

tual patterns to which they were responses. This was an astounding breakthrough—the key to all subsequent human progress—and likely precipitated what Donald (p.10) posits as the second cognitive transition, that of lexical invention. The concept of symbolizing, once grasped, must have exerted tremendous evolutionary pressures on hominids, at the very least favoring those whose brains had the plasticity to take full advantage of it, and possibly even being the primary determinant in the sudden appearance of the Cro-Magnon type. Moreover, the realization that representing objects and actions with symbols made it possible to invoke their stimulating effects even when those objects or actions were not actually present was an idea that would have spread meme-like into all other aspects of life. Once vocalized symbols made their appearance, the concept would have been quickly applied in other mediums, giving rise to painting and carving, body markings, dance, and ceremonial rituals. Tribal culture was born.

Consciousness was born then, too. By consciousness, of course, we mean access-consciousness—the skills involved in manipulating symbols that are representations of perceptual patterns. We mean, in fact, what in past debates has been called *thinking*, as in the centuries-old question, "is language necessary for thinking?" The answer being, yes, it is, but only in the sense that with language we learn the concept of symbolizing, a concept that can be applied in other than purely linguistic ways. Donald leaves the question dangling when he claims a pre-linguistic use of language, a "quasi-symbolic communication" (1997, p7) based on innate talent for miming, which sounds safe enough but actually obscures the issue by fuzzying the term *symbol*. Either something is a symbol or it is not, and what proto-humans did communicatively before realizing the value of symbols was *presymbolic*. Imitating, mimicking, gesturing, mugging, posturing, and miming can all effect changes in a social environment, but they do not use symbols. Certainly symbols are created out of such behaviors, but there is a line that behaviors cross where they cease implying and start symbolizing. And that is not an easy line for the mind to cross. It is a jump to a new level of cognitive processing—an epiphany understanding that surely spread like wildfire when Homo sapiens achieved it, dooming to quick extinction other hominids mentally incapable of grasping the concept. The consequence of this novel style of thinking would have been, analogously, like providing a primitive African tribe with firearms; in a short order there would be no other tribes. As soon there were no other hominids but our ancestors.

Historians, like philosophers and scientists, will sometimes see the trees and not the forest. They fail to recognize that human history is the history of consciousness, not of symbols per se. Symbols are but tools. The skills for employing them change, and thus is the course of history determined. Consciousness itself changes. That is why the cross-disciplinary work of theorists like Julian Jaynes and Merlin Donald is so important—for both history and psychology. We echo them here when we posit that the two watershed events in human history were the inventing of vocal language, and the inventing of means for writing language down. During the first of these events consciousness appeared, and during the second it was radically transformed—in what Donald (1997, p.13) calls the "externalization of memory." However, we would go much further than either Jaynes or Donald, because we are defining consciousness as *the modification of the state of awareness with skilled use of symbolic representations.* Why should there be only one technique for such modification? In fact, it is abundantly clear that conscious reality varies considerably from culture to culture and era to era, not to mention from age to chronological age. To deny the protean nature of consciousness is to deny the validity of what Whorf and Sapir, Luria, Piaget, Mead, and thousands of other ethnographers, anthropologists, and psychologists have taught us. It is not simply that we change which symbols we use. We actually change the way we use them. Techniques, skills, paradigms, *mediums* change. And, as they do, consciousness is also altered. Stages in the western development of written language—from cutting marks on clay tablets to record economic data, to carving ideograms in stone, to inscribing epics on scrolls, to manufacturing books by use of moveable type—can be correlated with different ways of thinking. Most noticeably, conceptions of identity have changed. Society has gradually lost all vestiges of tribal identity, and become an amalgam of autonomous individuals. There is every reason to believe that literacy has played a causative role in this, rather than being a coincidental process. In fact, if the evolution of consciousness is not taken into account, the course of human progress is inexplicable, and we can't even explain how consciousness appeared at all.

Jaynes, in his now-legendary book (Jaynes, 1976), audaciously argues that consciousness originated with a collapse of pre-literal societies that were organized very differently than our contemporary ones. His use of the term, however, makes it dependent on constructs of personal identity and ego, so he should have spoken of the origin of *modern* consciousness. We take the more defensible position here that in what Jaynes calls bicameral societies, there was an *earlier* form of consciousness. The tribe—and early city—functioned as a collective identity, and the manipulation of symbols took place openly, rather than in minds of individuals. The social environment was rich with symbols—vocal, gestured, artistic—passed back and forth in all sorts of complex feedback loops. One person responded to internal or external stimuli, using vocal sounds to reduce their effects, and the spoken words—being stimuli—evoked vocal responses from another person. These new stimuli cognitively overlaid the first person's original perception, triggering a new and modified response, which triggered a modified response in the second person. With third, fourth, and fifth persons in the mix, the buzzing interaction wove a tapestry that was *the* stream of consciousness. In this tribal culture, awareness was personal, while identity and consciousness were collective.

Individualized consciousness was yet many millennia away. It was not until the invention of literal mediums that human beings began learning to manipulate representational symbols internally. Silent speech became possible, clumsy at first but increasingly fluid and adept, until the slightest muscular tensions could be used to symbolize covertly what had previously been symbolized overtly. What had been societal feedback now was somatosensory feedback. Individual identities were hypostatized from the new internal tapestries, and the collective social identity disintegrated. Bicameral civilization collapsed, and empires were born. The modern era began, in an explosion of individualized consciousness sustained by the subtle machinery of self-manipulated muscular tensions.

No doubt this last statement may strike the reader as outlandish. What empirical evidence supports it?—is the demand. If consciousness were such a simple mechanism, surely it would have been deciphered a long time ago! But that is precisely our point. If awareness is understood as inherent in integrated systems like the human brain, then consciousness immediately becomes explicable. Because of the assumed confluence of the two terms, neither phenomenon has been seen for what it is. Once we factor systemic awareness out of the conflux, it becomes obvious that consciousness is the skill of employing behaviors representatively to layer perceptual patterns. During the bicameral period, this layering was done through interactions with other people. When it became possible to translate vocal symbols into abstract visual symbols, one person alone could accomplish the layering. It may have been laborious at first, with crude thoughts, but as the new skills were learned, and permeated society, human beings acquired a new and sophisticated form of consciousness, using almost imperceptible flickers of muscular tension that were no longer recognizable as remnants of speech.

As for empirical evidence, experimental psychology established long ago that directed thought and muscle tension were strongly associated (Humphrey, 1963). By the end of the 1930's, unanimity was reached on a few conclusions. Firstly, imagining or recalling an action induced minute tensions, particularly in muscles used to perform the action. Secondly, various types of mental tasks triggered specific muscle groups, with visualization inducing tensions in ocular muscles, arithmetic calculation inducing tensions in thigh muscles, and verbal tasks inducing tensions in the throat and jaws. Thirdly, total relaxation precluded directed thought. More recent experiments by Benjamin Libet (1985) indicate that consciousness lags a half-second behind action. Since that is an approximation of the human reaction time, it does not seem implausible to suggest that consciousness is more an echo of tensions in the body than the cause of them. The hypothesized order of events would be: stimulus, perception of the stimulus by the sense organs, excitation of cerebral neurons, activation of the motor cortex, efferent signals to muscles, tensions, proprioceptor afferent responses to the cerebellum, and, lastly, apperception by the cerebrum—consciousness.

In conclusion, we acknowledge that this interpretation of systemic awareness, thinking, and consciousness is an inevitable oversimplification. The subject of asymmetrical cerebral laterality has been wholly skipped, for example, although it has huge significance in the evolution and dynamics of consciousness. It was also necessary to assert that all behaviors are based on stimulus-avoidance, which, although fundamentally correct, does not present a complete picture of motivation. Nor did we explore the interplay between consciousness and systemic awareness, an interplay without which we wouldn't have the conscious experience. What we have attempted is to reinvent psychology by defining and distinguishing a few essential terms—absolutely and functionally—in relation to each other and to the Nothingness from which our reality emerges. It is our belief that the days of specialized and separate disciplines are waning, and that in the coming decades all of human investigations—psychology, history, physics, and the others—will converge, gradually at first, with cross-disciplinary work, and finally in a unification undreamed of since Aristotle, Pythagoras, and Plato. It is toward such a goal this paper is dedicated.

References:

Bell, J.S. (1964) On the Einstein-Podolsky-Rosen Paradox. [Physics] 1: 195-200

Block, N. (1995) On a confusion about a function of consciousness. [Behavioral and Brain Sciences] 18 (2): 227-287

Dennett, D.C. (1991) [Consciousness Explained] Little, Brown and Company

Donald, M. (1997) Precis of Origins of the Modern Mind: Three Stages in the Evolution of Culture and Cognition. [Behavioral and Brain Sciences] 16 (4): 737-791

Humphrey, G. (1963) [Thinking: An Introduction to Its Experimental Psychology] John Wiley & Sons, Inc.

Jaynes, J. (1976) [The Origin of Consciousness in the Breakdown of the Bicameral Mind] Houghton Mifflin Company

Levine, J. (1983) Materialism and qualia: the explanatory gap. [Pacific Philosophical Quarterly] 64:354-361

Libet, B. (1985) Unconscious cerebral initiative and the role of conscious will in voluntary action. [Behavioral and Brain Sciences] 8: 529-566

Pei, M. (1965) [The Story of Language] The New American Library, Inc.

Thomas, N. (1998) Imagination, Eliminativism, and the Pre-History of Consciousness. Paper delivered at the Toward a Science of Consciousness (Tucson III) Conference.

Tryon, E. (1973) Is the Universe a Quantum Fluctuation? [Nature] 246: 396-397

Unireality as a Metasystem

(2005)

$\mathbf{A}$ belief system provides context, and we can't make much sense of the world without one. By using a belief system, we can interpret the meanings of events, and choose appropriate responses. Since we find it necessary to use a belief system, we want one that effectively describes reality. There is no factor in human history more responsible for tragic outcomes than inaccurate and illogical belief systems. If we are to survive as a species we need to do better.

Everyone has a personal belief system—a self-context that explains who one is and how one fits into the world. The foundation for such belief systems is laid in early childhood, and evolves as one grows up and integrates new experiences into it. Rarely are such systems very systematic. Typically they are cobbled together out of half-truths, mistaken inferences, stereotypes, prejudices, and rationalizations. For the most part, however, they sustain our egos and keep us from flailing about in wild uncertainty.

But beyond these personal belief systems, we also have metasystems. These function as the glue of civilizations. They are far more rigorous and differentiated, having been hashed out during the evolution of culture, and laboriously formalized. In this general category belongs Christianity (and the other great religions), as well as Marxism and Science. Before these modern metasystems were others that even the experts only somewhat understand. The Mayan and Aztec cultures had powerful ones, as did the Greek and Roman and Egyptian civilizations. Metasystems have been with us since history began.

As explanatory paradigms, these metasystems have proved dangerous. They each focus on a single question about reality, at the expense of other questions. Everything is then related to that particular question; it becomes the only significant context. Christianity, for example, insists that the *why* of creation is what matters. Marxism argues that the *how* is the significant question. Science claims that everything is a matter of *what*. The metasystem then interprets all aspects of reality through the answer it has given to its own question, and brushes aside any other questions, which it perceives as threats to its legitimacy. It becomes all too easy for the Christians to torture heretics and witches, for Marxists to slaughter millions to hurry industrial progress, and for Scientists to perform ghastly experiments on the helpless. With the power of a state behind a metasystem, the repression of competing belief systems is a predictable outcome, not an aberration.

It is no small wonder that many people today are skeptical of any metasystem. Some of them express their aversion with cynicism and apathy, as if by denying loyalty they can somehow subvert the dominant metasystem. Others have turned to the spirituality of Zen, which eschews meanings, and proactively sidesteps the problem. This is at least an intellectually honest attitude. The most ardent metasystemist would do well to remember that frogs have survived successfully for a hundred ninety million years by just *being*—not caring what things meant.

Unfortunately, metasystems are an unavoidable consequence of using abstract symbols. If animals get can along fine without metasystems, it is only because they aren't endowed with the cognitive ability to consciously represent elements of reality with vocalizations. Development of language involves the construction of a *symbolic* reality—a conceptual tapestry representing true reality. As true reality is a unified whole, so symbolic reality must also be unified, and that can't be done without a metasystem.

We are not going to escape from metasystems, not as long as we employ symbols in the conscious way we do. It behooves us, then, to seek the metasystem that is least likely to provoke hostilities and intolerance—i.e., that applies to all of reality and doesn't need to repress contrary points-of-view. We should note here that doesn't necessarily make it *correct*. Epistemologically speaking, there probably isn't a *correct* way of describing reality in symbols. One way might be more accurate than another, or more useful, or more complete. But even at their best, symbols and their symbolic realities can only be approximations of true reality.

If we accept this caveat, and do set out to invent a better metasystem—one that treats all the questions about reality as being of equal relevance—we quickly discover the magnitude of the task. We must explain everything in a single paradigm. Not merely *why* we exist (as Christianity does with a god-concept), or *how* things change (as Marxism does with its dialectic materialism), or *what* causes what (as Science does with its physical laws), but also the questions of *when* and *where*, because time and space must be explained, as well. These separate questions merge into a single, ultimate question—what relationship is there between Nothingness and Somethingness? It is there any metasystem must begin, or all its grand theories are suspended from skyhooks. Until that can be answered, symbols can only be defined in terms of other symbols; there is no solid footing anywhere. It is therefore the quintessential problem of the curious mind, and the first test of any aspiring metasystem.

If we put our familiar metasystems to this test, they fail it miserably. Christianity claims that a divine being created the world. Any subsequent questions—as to what created this creator, for example—are peremptorily rebuffed on grounds of ineffability. But it is logically impossible to build a logically consistent system of meaning on that which logic cannot address. We may as well throw up our hands if that is the case, and forget using logic, because we never will make sense of things. All the theologians toiled in vain.

Marxism pretends no interest in how the world started. In human history it discovers the law of the dialectic, and grandly flourishes it like an edict from the emperor himself. But does it apply to quantum physics? Does dialectic materialism explain the complexity of the ecosystem? What about its relevance to fine art, or Mozart? The silence is deafening.

Science accepts the assignment of logically relating Somethingness to Nothingness, but then stumbles. The universe, says the scientist, began as an explosion of matter and anti-matter in equal quantities, most of which was mutually annihilated. But there was slightly more matter in the equation, and that accounts for the universe we now see. Why was there that imbalance at the start? The scientist isn't sure. But he is sure he can explain everything that happens after that first nanosecond. Which means that when all the esoteric rhapsodizing about branes and multiverses is cut short, our question still remains.

One reason our familiar metasystems stumble on this question is that they have all taken for granted that there once was Nothingness, and now there is Somethingness. But this is clearly a fallacious assumption. Time flows in Somethingness, but not in a Nothingness. Nothingness has no characteristics, no dynamics, no features. Before and after are meaningless terms when one is talking about Nothingness. It is also inaccurate to say that Somethingness exists at the same *time* as Nothingness, since time exists only within the Somethingness. There seems to be only one way to relate the two, and that is to consider them identities. That is, Somethingness is just a particular way of looking at Nothingness. This answer elegantly suffices to get us through the temporal part of the problem.

As to why Somethingness comes out of Nothingness—which is the teleological part of the problem, and over which dreadful wars have been waged in the name of one god or another—we can brush the question aside. It is erroneous to assume that causation applies to the manifestation of our Somethingness, because causality is, like time, a feature of Somethingness. Causality does not apply to Nothingness. In the context of Nothingness, it has utterly no meaning. Where time does not apply neither does causality, because a length of time must separate cause from effect. Hence, a Nothingness that is devoid of time requires no causal agent to turn it into a Somethingness.

Nothingness, in fact, has infinite potential. It contains every possibility. Our universe is certainly not the only whim it indulges. There is one limitation on the cornucopia—any universe it makes must be zero-sum. Which is to say, it must amount to nothing, like ours. Many of them may be virtual likenesses of ours—so-called parallel universes, with altered outcomes. Germany might have triumphed in World War II, or Lincoln not been assassinated, or dinosaurs might not have gone extinct. The Copenhagen Interpretation of quantum physics is that a parallel universe appears every single time the smallest choice is made by anyone—or anything. In that scenario, Nothingness is like a great hall of mirrors with every image slightly different. But why limit the Nothingness to universes like ours? Infinity is infinity, and other universes may be so unlike our own as to be not only unfamiliar but also uninhabitable. Planck's constant or pi might have different values in such universes. They could have six dimensions, or twenty-six. They might have more than one kind of time. As long as the potential universe amounts to zero, it can't be rejected.

Logically, there are two inferences we can draw from this. The first is that the features of our universe are like terms in a zero-sum equation, and must have the specific values that they do. If pi had a different value, or if there were another spatial dimension, or if electrons were heavier, or if gravity were stronger, then our universe would not exist, and neither would we. All of it fits together precisely.

The second inference is that whatever occurs in our universe, the zero-sum is maintained. Our universe must start at the beginning as a zero-sum, but it must continue as one. Nothingness does not care how many billion years have passed in the Somethingness; it cares not a whit. From its point-of-view, our universe is continuously re-emerging. At every moment it must be equal to zero. Therefore, with every action, the universe must do whatever is necessary to re-balance the equation. The simplest example of this balancing act was succinctly expressed by Isaac Newton: for every action there is an equal and opposite reaction. Another example is the conservation of mass/energy—it can be neither created nor destroyed. It is this same requirement of maintaining a zero-sum that is responsible for the evolution of form itself in our universe.

We have argued that our familiar metasystems can't provide a bridge between Nothingness and Somethingness. Christianity (like other religions) mounts a defense of ineffability to escape the question, Marxism pretends the question is irrelevant, and Science works backward only to the instant after Somethingness appeared. But that bridge is essential. Without it we cannot trust any metasystem, and if we resign ourselves to untrustworthy metasystems we must also resign ourselves to never knowing anything for sure.

This is far more than just an existential problem. People need a metasystem with which they can organize their thoughts and feelings, and if that metasystem is a flawed model of true reality they will surely wander down perilous paths. Worse still, a flawed metasystem doesn't work in the real world, no matter how neat it looks on paper or how eloquent the rhetoric used to promote it. When it fails to generate anticipated results, a true believer inevitably redoubles efforts rather than abandoning the flawed model—which means more torture, execution, genocide, and war. History reveals this process played out again and again, most memorably with the Spanish Inquisition, the Third Reich, the Reign of Terror, Bolshevism, and the Cultural Revolution. It will be played out again and again, with increasingly global consequences, if we cannot discover an anchored and trustworthy metasystem.

Such a metasystem would begin with the fact that our universe must be a zero-sum. At first blush it might seem there is no way that something can be made out of nothing. If we look at the world around us, however, we see a great deal of nothing. We see mountains and valleys, wet and dry, north and south. We see males and females, tall and short, bold and shy, left-handed and right-handed. We see future and past, up and down, forward and back, positive and negative. In other words, we see pairs of opposite terms that—as in the arithmetic of -1 and +1—add up to a sum of zero. When any one of these pairs is combined, the terms cancel each other out. Add together all the mountains and valleys of the world, and you have an utterly flat surface; height and depth cease to exist. Add together the male and female, and the result is a neuter—an absence of gender. Add together future and past, and there is only now. So it is possible to make something out of nothing. It only requires that the nothing be separated into a positive and negative that can cancel each other out. With opposite terms, a zero-sum universe can be constructed.

The most common misapprehension of everyday life is that we are surrounded by things, and that those things are real—concrete, tangible, and constant. Until the twentieth century there was no reason to doubt it. Science has learned that it is illusion. Things, when they are analyzed down to their smallest parts, cease to be solid at all. They are made entirely of fields that are just like the gravitational and magnetic fields we are familiar with. It is true that physicists still refer to the tiniest parts of things as particles, but they don't mean the term literally. The electron, for example, is a well-studied particle. Yet as far as anyone can tell it has no size. The electron is a point with properties of charge and spin and mass. And it doesn't revolve around the nucleus of the atom like earth does around the sun. It is a standing wave that encircles the nucleus. And so with other sub-atomic particles. The fact is that we are surrounded by things that aren't real in the conventional meaning of the word. Reality is a weave of invisible fields. Solidity is an illusion.

The curious thing about these fields is that at least some of them are opposite terms. No physicists have yet dared to suggest that noth-ing exists except fields of equal and opposite values, but the conclu-sion is there to be drawn. And we will indeed draw it. We will follow logic here—that the only way to reconcile the terms of Noth-ingness and Somethingness is to interpret them as identities, that the Somethingness of the universe must therefore be a zero-sum, that the only way to construct it is out of opposite terms, and that those terms manifest as fields.

There are four fundamental forces, so there must be four pairs of opposites, each with a positive and a negative aspect. Physicists call these four forces the strong, the weak, the electromagnetic, and the gravitational. Of the strong, we know only one of its aspects, the one that binds nucleons together in the atom. The weak force, as revealed in beta decay, is associated with neutrinos, and these have either a clockwise or counterclockwise spin, spin itself being a field. The gravitational force has two aspects, the one pulling matter together, the other pulling it apart with expansion of the universe. And, the electromagnetic force has well-known positive and negative aspects. Thus, there are four types of opposite terms that make up our uni-verse.

It may seem an astonishing claim, that reality is fashioned out of nothing. Shouldn't someone have noticed? But that which is obvious is often overlooked. Perception is less straightforward than we think. This doesn't mean simply that one sees what one expects to see, or that the eye is not as quick as the hand. It means that perception is flawed by its very definition. If, as we argue here, the universe is made out of opposite terms, then we cannot directly see that it is Nothingness. To see plainly both the positive and negative values of the terms would be to see nothing. We are only able to see a repre-sentation of the Nothingness.

But that shouldn't seem surprising. After all, *anything* we per-ceive is representational. With our eyes we see a two-dimensional representation of a three-dimensional tableau—a flat-screen image on the retinas. If we could somehow see in three dimensions, we would see the backs and sides of things and not just the fronts facing us. We think we perceive in three spatial dimensions only because our brains learned early in life to use visual cues of shading and perspective and binocular images to suggest depth to reality.

Similarly with any other kind of perception. Each sense organ is but a two-dimensional surface on which four-dimensional events are represented as shifting patterns. These patterns are broken down into (zero-dimension) bits of information, sent to the brain as one-dimensional streams of information, and the brain reconstructs them into maps of three-dimensional reality. In short, we never perceive actual reality. We perceive representations of it—patterns sketched in fewer than three dimensions.

But three-dimensional reality is itself also a representation. It is but an image sketched in three dimensions of a Nothingness subdivided into four dimensions. That is what the pairs of opposite terms are—dimensions. But they don't fully manifest themselves. They manifest only as a three-dimensional representation. The universe we consider real is but a three-dimensional image.

The image is a flawlessly accurate one. With an analogy we can demonstrate what that means. The paintings of ancient Egypt are not accurate images. The pharaohs and slaves seem flat, unrealistic because their figures haven't got depth. There's no hint that their world had more than two dimensions. It might have been Flatland. It's true that a painting can't have real depth; it is, after all, a two-dimensional depiction. But it can have implied depth, which is how we paint today. Contemporary art conveys depth with techniques of shading and perspective. The implied depth makes paintings more accurate representations, because scenes that were painted *did* in fact have three dimensions. Which means that if one is making a representation in fewer dimensions than the original (which is, in fact, how one makes a representation), one needs to show the omitted dimension in some indirect way if the representation is to look realistic—that is, accurate. If our universe is to accurately represent the Nothingness subdivided into four dimensions, it must imply the omitted dimension. That omitted dimension is called time, and it is expressed with form.

What is form? Stars and galaxies are form, atoms are form, we are form. Whatever is apparently real is form. As we noted above, when

114

one attempts to break this form down into its constituent parts, one discovers it is composed of tiny interlocked fields. These fields are the equivalent of shading in an art painting. They are bent and curving evidences that four dimensions are being shaped into the three-dimensional image of reality. What we see, when we look at reality, is the geometry of those curves—and that is all. We see a three-dimensional matrix of fields bent into form.

What is especially interesting about this matrix is that all four of its original dimensions are expanding. Space is expanding, and time is going forward (that is, expanding) as well. Thus, along with expansion of the three spatial dimensions, form must be continuously evolving as well, for the representation to remain accurate. Bending and curving steadily increases; the convolution of form deepens. In this sense, form represents the amount of passed time, just as the spatial size of the universe represents the amount of expansion in the other three dimensions. As space expands, time goes forward, and form evolves.

Here we might pause and consider how radical an interpretation of reality this is. Firstly, while modern physics acknowledges that the universe might be a "vacuum fluctuation," it makes no claim that our universe is zero-sum. Conventional thinking is that for our universe to be zero-sum, there would have to be an equivalent amount of anti-matter for the matter in it. Secondly, it is heretical to suggest the four fundamental forces have positive and negative values. Magnetism clearly does, but there is no evidence of a negative strong force, and spin is not considered a force at all, even though it is closely associated with the weak force. Thirdly, physics thinks of form as merely an incidental effect of the Big Bang. Form is predicted to decay as the universe expands, rather than evolve. Evolution of form seems to violate the Second Law of Thermodynamics. So, our description of reality is too radical for physicists to accept.

However, it is not so inconsistent with physics as to be judged implausible. We wouldn't trust a metasystem that was physically implausible, because the laws that have been formulated by physicists are ultimate reductions of the processes of reality. However, we haven't contradicted those laws here (with the possible exception of the Second Law of Thermodynamics). Rather, we have tried to tie them together in one holistic framework that bridges the chasm between

Somethingness and Nothingness. By doing so, we have challenged what is called the Standard Model of physics, but that is not a set of laws, only an interpretation of how the laws might fit together. Although the Standard Model is the conventional story about how the universe came to be what it is today, it is demonstrably incomplete. Physicists cling to it not because it is complete, but because there are no worthy alternatives. Our interpretation of reality as a three-dimensional image of Nothingness is at least as consistent with the laws of physics as the Standard Model is, and answers many of the questions that model leaves open. As such, it provides the foundation for a new metasystem: unirealism.

On Intimacy

(2006)

I have come to believe that the quantum phenomenon of correlation plays a profound role in how human beings relate to each other and form bonds and sustain them. In quantum correlation, two or more particles, when they interact in some way, surrender their isolate identities, with a blurring of the ownership of characteristics. This doesn't mean that the individual traits are somehow divided equally between two particles, for the traits are in fact quantized and indivisible. One can't, for example, split the value of spin in half and give half to each interacting particle. Rather, the spin (or momentum, or polarization, or other observable characteristic) ceases to belong exclusively to the particle that originally possessed it, and begins—suddenly—to belong to the pair, as if the separate particles are now conjoined, or entangled. The pair becomes a unified entity with the characteristics of its constituent particles, no matter what distance separates the particles.

This is consistent with the Biblical pronouncement about marriage that "the two shall be as one." It is what we mean when we refer to an intimate relationship in the romantic sense. The married couple, in the ideal, possesses an invisible identity that is something more than the sum of the husband and wife's identities. It is perhaps tempting to think of this conjoined identity as a hypostatization—something fanciful, real only to dreamy-eyed lovers, an emotional chimera. The correlation phenomenon in quantum physics also seems fantastic. Einstein, in fact, argued that the logical implications of correlation were so absurd that they proved quantum theory—which he disliked—incomplete. Fanciful as quantum correlation sounds to a realist, however, it has been confirmed in experiments in physics labs. It seems to be a property of human relationships, as well.

We should note that correlation in human affairs is not limited to marriage, or just romance. As with subatomic particles at a quantum level, any kind of human interaction can quite easily become correlative. When a pair of individuals, or a group, take on an identity greater than the sum of their parts, it is because a correlation is formed. This happens quite often with teams working or playing together, although the phenomenon tends to be fleeting. Correlation explains also why mobs form, and behave as they do. In this essay, however, I am only concerned with the role correlation plays in intimate relationships, especially long-term couplings of the marital type.

In the quantum realm, interaction is what causes a correlation to form, and the making of an observation is what breaks a correlation. Physicists usually think of an observation as being some measurement. Two correlated particles may share their respective values of spin, for example, but the moment that the researcher checks either particle to distinguish which kind of spin it has, then the shared values separate, and each particle gets assigned one of the spin values. With that observation, the quantum wave function that represents the entangled characteristic collapses and it localizes. How and why correlations occur, and later break, is a mystery to physicists. There has also been debate for a century as to the manner in which a break in correlation translates into an event in our everyday reality, but we need explore that debate here. We can understand intimate relationships as being connections of quantum correlation, and turn our attention to the more important problem of why such correlations break. Why does love disappear? Why do promising marriages end so often in divorce? For if intimacy is a matter of correlation, then it is the breaking of correlation that ends the intimacy, and such a breaking is always a consequence of an observation.

An observation in this sense always indicates some measurement that compels that which is being observed to stop its identity sharing, and be only what it is. This is an important point. Physicists take the attitude that they are directly breaking correlations by their acts of measurement, but what they are actually doing is forcing the correlated particles into a situation that compels them to identify themselves. The physicist himself is not breaking the correlation. He is instead setting up for the particles a stressful situation within which it is impossible for them to remain correlated.

In the same way, circumstances in an evolving intimate relationship often put such great stresses on a couple that one or both of the spouses feel compelled to identify themselves as individuals. If this happens, the correlation breaks. The magical shared identity then vanishes, leaving two individuals connected only by tangible evidences of marriage. The difference between correlated subatomic particles and correlated partners is that particles don't have free will. The researcher can lead them around the lab by their noses, making them correlate at his command and then, at his whim, break the correlation. The married partners, on the other hand, can opt to save their correlation. They can refuse to make the kinds of observations that turn them into individuals. To avoid those observations, and to preserve the connection of intimacy, can be sometimes difficult, but only in the direst exigencies is it impossible. This point is made clearly in *1984*, a novel unforgettable not only for its chilling political slogans, but also for the dreadful measures employed by the State to destroy the intimacy between Winston Smith and his lover. That such drastic measures are required to overcome the free will choice of maintaining correlation should demonstrate that when we allow correlations to break, it isn't because we are truly forced to, but rather because we let it happen.

Then why does it happen so frequently? I think the easiest answer to this question is that the ego is too quick to defend itself. Ego must be suspended, at least within the bounds of the marriage union, if correlation is to continue. During the initial stage of a romance, the ego is happily subordinated, but as the biochemical thrill we like to call romantic love wears off, the ego re-asserts itself. The human psyche is a veritable minefield of prides, fears, and insecurities. Accidental or careless slights by a partner can trip a waiting mine, as can the opinions of one's family or peer group. The ego, no longer sedated as it is at the beginning of an intimate relationship, rises to defend itself. It makes its first, but not its last, observation of the other partner. It would be kinder if we could say that this first pulling-away is imperceptible, but I don't think that's true. Rather, it is like the first step down the road to corruption, and only the most callous human being could fail to recognize the import of the choice. It is easily rationalized, however, and the next steps become easier and less shocking.

We can be quite specific about the sort of observations that break correlated relationships. Invariably, they begin as acts of defense, and proceed to a negative naming of the partner. The negative naming is what does the damage. Strangely, or perhaps not, being wounded or made fearful by the word or deed of a partner does not in and of itself break the state of correlation. Nor are all responses correlation-breaking. But it is also true that negative naming of the partner is just as damaging even if it is only thought, and not spoken aloud. If a husband thinks to himself, in reaction to his wife's behaviors, "She is so selfish!" he is separating himself from her just as effectively as if he voiced the complaint to buddies at the tavern. A correlation is an intangible thing. It is broken just as thoroughly by a silent griping as it is by a public denunciation.

Does this mean that one can never safely talk to or about one's partner? Not at all. It is possible to name one's partner—silently or aloud—without fracturing the correlation. One can say, for example, "My wife is worried, that is why she is harrassing me." Alternatively, the husband can even say to his buddies, "I dearly love my wife, but sometimes she drives me crazy." That statement is not a renunciation of the intimate correlation, but the remark can serve to mitigate the frustration or pain that otherwise might be expressed as a negative naming. Of course, it is also safe to name one's partner in a positive way, as in, "She's the kindest person I've ever known." Generally speaking, no harm is done by the non-defensive describing of a part-ner. Negative naming, on the other hand, is purposefully defensive. It seeks to package the other partner with a label so that he or she may be held at arm's length. Negative naming is an undeniable effort to break the correlative state, and although it is done impulsively, with-out any special skill, it is effective at its purpose.

These distinctions notwithstanding, it is probably wisest to make a habit of never naming the intimate partner for any reason. A bit of reflection informs us how infrequent are those occasions when it is really necessary to assign descriptive names to a partner. It is much better to stay *in* the relationship at all times, never stepping back, never separating. When called upon by some third party to describe one's partner, reluctance should rule any answers. Better to say noth-ing at all than chance breaking that intimate connection.

It may be unfair to say this, but I do believe that of the two sexes, women are more likely to break correlations, not because they are more careless about the value of intimacy, but rather because they are far more verbally oriented than men are. Men do not generally speak about their wives—positively or negatively—to others, and they are less given to think in words. On the other hand, it seems to me that, of the two, it is men who are more prone to behaviors that incur the negative naming that breaks correlations. Men express their feelings more violently, and are somewhat less conscientious about sexual fidelity, and it is surely true that abuse and adultery are the most devastating circumstances to challenge an intimate relationship.

Ultimately, every couple must cope at least from time to time with stressful situations in their relationship, and can do extraordinary damage to it with a single thought. The key to preserving the intimacy is quite simple, really. It is to choose intimacy, whenever the choice is presented. Every time it is not chosen, the bond of correlation is broken. It may be that once broken, full correlation can never be restored. It may be that correlation can bloom again, with patience and commitment and continued interaction. But it is certainly the case that repeated breaking of the correlation has a cumulative destructiveness, and the point is eventually reached where it can never be regained. Too many marriages are lost this way, too much bitterness and resentment etched forever in the hearts of partners who once loved. It needn't be that way, but until lovers understand that their love has a basis in quantum reality only the most fortunate of them will know how to take care of it.

* * * * *

Part III

Theory of Unireality

----One----

Grana: What is the theory of unireality?

Heidrich: The theory of unireality is an explanation of what things are and how they change.

G: And what does it say?

H: That the universe is a zero-sum equation. That our universe is constructed out of four imaginary binaries, each binary being a zero-sum in itself. Therefore, perception of our universe is only possible from an internal point-of-view. Which is to say that if the observer is a complex of these four-dimensional values, he will perceive a universe around him. What he sees is not true reality, but a representative model of true reality. He sees a symbolic reality. He is, in fact, a function of that symbolic reality. The theory goes on to state that given this zero-sum condition and given the original blueprint for making three-dimensional symbols out of four overall dimensions, all change in the universe must follow a single set of transformations.

G: You've lost me already.

H: The theory of unireality says that the universe adds up to nothing.

G: My senses are deceiving me? The universe isn't there at all?

H: Not objectively. It exists only because we're inside, observing it.

G: But how can it be real on the inside, and not real from the outside?

H: Because we observers are part of the illusion of its existence.

G: Then there is no reality. It's all illusion.

H: No, no. Since you yourself are part of the illusion, the illusion is very real indeed.

G: So the illusion is a subjective experience, while the objective reality is Nothingness.

H: Exactly.

G: Assuming I were in a spaceship and traveled very fast, for a long time, and came to the edge of the universe... and somehow broke through and got outside, I'd see nothing at all behind me?

H: Yes. But you couldn't do that.

G: Why?

H: You're an integral part of the illusion, and you can't get outside it. You'd merely drag it with you when you tried to break through. You'd drag mass and spacetime with you. And once you got outside, it would no longer be the outside.

G: Then we can't get outside to discover the inside doesn't objectively exist. So how can it be proved what you're saying is true?

H: I suppose you won't take my word for it.

G: For the sake of argument.

H: Then let's get into your spaceship and take a journey to the edge of the universe. Remember, the furthest boundaries of the universe are receding from us at more than 90% the speed of light. To get to the edge would require that your spaceship travel even faster. Shall we say at 99.99% that of light? To reach such a speed requires a tremendous fuel supply, because the faster you go, the more the energy you burn is converted into massiveness of the spaceship. To reach the speed of light would require an infinite supply of fuel, and at that speed your spaceship would have infinite mass. And at that speed, of course, relative time will have stopped. From the point-of-view of any loved ones you left back home, you on that spaceship will have ceased aging. To them, your clocks will have stopped.

G: Okay.

H: To reach the speed of light and have infinite mass, you've had to burn up all the mass of the universe as fuel, and as you keep glancing back the last remaining clock is going very fast. At that magic moment when you have just expended your last atom of fuel—you've even tossed the last clock into the drive—you have also reached the speed of light, and become infinitely massive, and at that very instant you cross the edge of the universe. Look back now, what do you see?

G: I certainly don't know.

H: You can't see any mass; you've burned it all. Space? Space is a function of fields, of which none will remain, because the ex-universe has no matter left in it to generate fields. Time? There's no time passing back where you came from, because your own time—relative to what the rate of time back home was—has infinitely slowed down. Which means that time back there has infinitely speeded up. In the mere blink of an eye, the entire ex-universe is born, expands and collapses. So quickly you won't even see it occurring.

G: So...

H: So when you look back you'll see no mass, no time, no space. You'll see Nothingness. The ex-universe won't exist, objectively.

G: In other words, this universe we've apparently got, it's something from nothing. It doesn't exist, but it seems to, because we're part of it. And you claim it's constructed out of four imaginary binaries?

H: Do you understand what is meant by a binary?

G: A binary is a set, such as yes/no, or life/death, or right/left.

H: Yes. There are two equal and opposite terms in a binary. When you sum them together you get absolutely nothing. There isn't much you can say about the sum of a binary set, except that it isn't there. Objectively the sum is Nothingness. To express it another way, a binary is a differentiation, and when it is summed it is a non-differentiation. With a three-dimensional figure, a cube, for example, dimensions are defined by three binaries: up/down, which tells height, forward/back, which tells depth, and right/left, which tells width. Take away any of those binaries, and you've erased a dimension. Up and down add together to make a sum of nothing, so to not make a differentiation as to height is not to say that there is no height, but only that you're disregarding height. You might say that the sum is neutral about the characteristic which the binary proposed to distinguish. The sum of up and down becomes a complete neutrality about height. With the cube, if you get rid of the up/down binary, you have a geometric figure in two dimensions. Height is unspecified.

G: A binary, then, is a zero-sum entity.

H: And no matter how many zero-sum entities you have, the total is still Nothingness, correct?

G: Yes.

H: Each of these binaries has a positive and a negative value—equal and opposite terms. How do you know which is positive and which negative? You don't. Positive and negative must be distinguished by context. So let's consider the four imaginary binaries out of which we construct a universe. Let us take one binary and use it to link together those ends of the other three binaries that we arbitrarily label as positive. Then we'll always know which is positive and which negative. As with positive and negative electric charge: the positive is that found on a proton, which is a linkage of positive ends.

G: From this arrangement of binary sets comes a universe?

H: Yes. In the instant of linkage, you've started the flow of time and the expansion of space. It's easy to see why. This linking is, in effect, a representing of four values in three. The Nothingness is bent into a negative three-dimensional space. The negative values will curve in this bent spatial matrix, the curves will become cycles, and time will move forward as space expands to accommodate the developing cycles.

G: And you've gotten something from nothing.

H: Really, you haven't. It's just that all you can see is negative space and what is transpiring in that negative space. It only looks like something because you're selectively perceiving binary values. In fact, there must be as well a positive space, filled with cycles developing oppositely, where time's arrow is pointed in the other direction. That mirror universe is filled with anti-matter. If you were to step outside and look back—if you could—you'd see nothing, because the two universes are equal and opposite. These two binary universes are what I call simulspatial. They are superimposed. This is evident from the fact that you can create anti-matter—with, of course, an equal amount of matter—in the laboratory, simply by focusing kinetic energy. The universes are equal and opposite, so there's no violation of the zero-sum condition in their existence. And each universe by itself can be cancelled down to zero. After all, just because we're in negative space doesn't mean that we're failing to see the entirety of the four binaries. We're simply seeing them differentiated with an arbitrary distinction of positive and negative.

G: And that distinction provides us with a three-dimensional matrix against which events occur.

H: Yes. You're treating one binary differently than the other three, in using it as a link. Which means, by the way, that you'll only perceive one of its equal and opposite values, in time going forward. Three binaries are being employed to represent four, and that fourth becomes time. Time is then continuously re-entering the universe with the bending, and revealing itself as cyclic motion, which itself can be broken down into dimensional components: waves, vibrations, and oscillations. So it isn't that the fourth dimension has vanished; it's there and visible, but expressed differently. The first three are expressed as spatial dimensions, while time expresses itself by what occurs in those three spatial dimensions. What is significant about this blueprint is the inference that the value of time-gone-by must always be included in the form of the universe at any given moment. Since the value of time-gone-by is expressed in cycles, this indicates that the more time that passes, the more complexly interwoven the cycles must become. In short, time can only pass if form evolves.

G: But at any given instant, we see a solid three-dimensional world.

H: Yes, because at any given instant you are disregarding time. Of course, when you "see" this world, it is registering on your retinal screens in two dimensions. You only *infer* that it exists in three dimensions, and with your hands you can touch and verify the presence of three dimensions. When you "see" a cube, for example, you're seeing a two-dimensional image of the cube, but you immediately notice the slant of edges, and the shading of perspective. The evidence of a third dimension is still there, hidden in the form. In similar fashion, time is concealed in symbolic form. Though rather than slanting lines of perspective, you see curving. And everything in our world curves, no matter how straight it might appear to be.

G: If I understand this correctly, our three-dimensional world is the equivalent of a photograph. In the former time is implied, so that four dimensions appear as three, and in the latter depth is implied, so that three dimensions appear as two.

H: Excellent analogy. When you look at the photograph, you realize that depth is still represented; it's just been expressed by modifying values in the other two dimensions. Perspective, proportions, shading—these tell you all you need to know about the depth.

G: And time is expressed in the three-dimensional image of solid reality by cycles—the bending in motion and repetition in events.

H: And what do cycles make? They make form. Hence form is the way a particular value of time expresses itself. And the more time that needs to be expressed, the more complex the form.

G: Ah! So, change is the process by which time becomes expressed!

H: Yes. Form is made up of curves—cycles and waves and vibrations—which are dimensionally balanced to provide apparent equilibrium. Form must change over time, because it must continuously integrate the increments of time into itself. Time as it passes *enters* form. This process can be understood if you think about what happens as you watch time pass. Your eyes are continuously absorbing time with the impact of photons on your retinas. These tiny entrances of time into form are "observations." An important term. As time passes observations are made. And, as observations are made, form changes. It changes incrementally, by the acquisition of symbols.

G: The photons I absorb become experiential information, which I convert into symbolic knowledge. Is that what you mean?

H: That's one instance. Bear in mind also, the number of symbols in symbolic reality doesn't necessarily have to increase geometrically, because there's also a progressive synthesizing going on, whereby symbols combine to make more sophisticated symbols. Rather like the fusion of lighter atoms into heavier atoms. So we're talking about several processes here: observing, symbolizing, and synthesizing.

G: And these ongoing processes are part of the single set of symbolic transformations you mentioned at the outset?

H: Yes. And they follow inevitably from the basic 4/3 reduction. Four binaries being represented by three spatial dimensions, through which time appears to flow. It is this 4/3 representation which gives us our illusion of being. It is the fundamental structure of symbolic reality.

----Two----

Grana: Why do we need a new theory about reality, when we already know so much about how it all works?

Heidrich: From the number of great unanswered questions, it's evident that knowledge is utterly fragmented.

G: Do you mean to say that all the great unanswered questions can be satisfied by a single theoretical answer?

H: They can all be answered by knowing how things change, yes. After all, what's our world made up of?—things and change. Isn't this true? Our very syntax reflects this. How do you write a sentence?—with subject and verb. But why? Why is it that reality divides so neatly into things and change? Why is it that we can describe reality in the format of subject and verb? And why does virtually every language devised by the human mind employ that basic syntax? It's very curious, isn't it?

G: Perhaps. But most of us get along admirably well without wondering about it.

H: Well, if you just take the format for granted, you do manage to survive and procreate, which is strong evidence that the distinction has some elementary significance. However, the larger questions remain, and I doubt there's anyone out there who hasn't occasionally considered what we really are, why we're here, and what it all means. But you cannot understand reality-as-a-whole until you understand the relationship of things and change, because there must be some fundamental reason why reality so readily organizes into things and change, subjects and verbs, objects and actions, being and becoming.

G: There are things because there are atoms, and there is change because those atoms interact. Or so they tell me. Personally, I've never seen an atom.

H: Ah, atoms and interactions in an expanding universe, gradually entropizing so that in the far distant future everything will be cold and dead. That's the conventional world-view. Nihilistic, cynical, but also an abject failure, because it can't even answer the most essential question in the physics of it. You know which one I mean?

G: My physics is a bit rusty. I recall that there was a dispute over whether light was composed of particles or waves. Maybe that's been resolved, I don't recall.

H: It hasn't. Physicists have compromised by saying it is both, depending on how you measure it, and ceased to fret about the problem. But it remains nevertheless an essential riddle, and perhaps the only one. Likely if you know why light manifests as both particles and waves, you go a long way toward solving all the nagging enigmas of physics. Because what are particles?—they're things. And what are interactions?—they're changes. And what, then, are waves?—they are things-in-the-process-of-change. I asked why there are things and change, and basically you replied that it's because there are things and change. You've told me nothing.

G: Perhaps it's a problem for philosophers instead of physicists.

H: And what is the essential question of philosophy?

G: I'm quite certain it has to do with Platonic forms.

H: Yes, what is the relationship between "universals" and "particulars?" The question was originally posed in Ancient Greece, but it has been debated ever since by men like Avicenna, Aquinas, Locke, Marx. And never satisfactorily answered. Basically, it's the question of connection between conceptual reality and physical reality.

G: Sort of the Cartesian mind/body problem.

H: That's another version of it. Let me ask you one more. What was the essential question of early Christianity? And still is, for all I know.

G: There have been so many disputes in the history of Christianity—such as whether full immersion was necessary for baptism—that I can't say one is essential.

H: The original problem, until the Catholic Church gained sufficient power to stamp out such heresies, was this: are people able to under-stand God directly, or do they need an intermediary? Obviously the church believes the latter. If people could apprehend God's will directly, nobody would get to dress up in fancy robes and expensive jewels. By the way, the clerics never quite managed to bury this question. For a long time they were successful, because nobody else could read, so all they had to do was insist that the gospels were God's word here on earth, and that gave them the job of reading and interpreting God for the common folk. When bookmaking was mechanized, and literacy spread across Europe, the pedestal from which the clergy pontificated vanished. Suddenly the layman could read the gospels for himself, so the essential problem resurfaced. Even if laymen couldn't apprehend God directly--which the church maintained they couldn't—they could certainly read the gospels and discover the word of God without priest and pope.

G: Seems to me you're playing fast and loose with history. The Reformation was initiated not by Gnostics, but by churchmen like Luther who wanted to do away with abuses of power by the clergy.

H: If the church had not originally staked out its position against *gnosis*, which is the idea that divine knowledge is inherent in human beings, then it never would have had the power to abuse in the first place. In the second place, the invention of the printing press—or, I should say, the application of the technology—was simultaneous with the burgeoning abuses of power for the simple reason that they had a common cause. That cause being the expansion of commerce and trade, which also birthed the Renaissance, and urbanization, and spread epidemic disease, as well as a host of other consequences. But the point I'm making is that in the nascent church the essential problem concerned this necessity for clerical intermediaries between God and man. The reformers, it is true, focused on the proper role of the intermediary, and not on whether such a role was justified to begin with. But when Luther posted his complaints he was saying, "Hey, I'm reading God's word here, and I don't find anything about purchasing forgiveness!" So he raised the lid on the box that contained the essential problem.

G: Okay.

H: So here are the essential questions in physics, philosophy, and the Christian religion. How can light be a particle and wave simultaneously? How can we reconcile the world of ideas with the world of sense impression? And, can ordinary people experience God's will directly? All of these reduce to the same problem, and this is, what is the relationship between things and change?

G: You've lost me on that.

H: The physicist Edwin Schrodinger had a cat. A make-believe cat. The cat is now famous, or infamous, because of a thought-problem Schrodinger used it in. Schrodinger had begun formulating descriptions of subatomic particles as probability functions. Quantum wave functions. Now, a lot of laypeople are highly distressed at the thought of even trying to grasp quantum physics, but it's very simple. When you get down to the tiniest particles of reality, they aren't solid and fixed, they're kind of uncertain. Put a whole lot of them together, and what you get is solid and fixed, but each single particle is sort of blurry. It isn't really anywhere in particular. Instead, it's spread out over all the possibilities of where it might be, given the location and form of the larger object that contains it. Simple. Not very commonsensical, but simple. Very esoteric math goes along with quantum mechanics, but it's irrelevant to the problem of Schrodinger's cat. Schrodinger, as I said, was describing these little particles by quantum wave functions. By sets of probabilities. The most interesting thing about these little particles is that if you perform the right experiment, you *can* nail them down. You can "find" them. The blur suddenly comes into sharp focus, and voila!—there's the particle. That nailing down is an observation.

G: Oh, yes, I recall. What Schrodinger did was set up a fictional experiment where his cat was in a box, concealed from view. There was a poison gas capsule, and by some device the particle was able to release the gas and poison the cat.

H: Yes. Remember, the particle isn't really a particle until it's observed. Until then it's only a wave function. When it is observed, the wave function collapses, and the particle appears, and at that point triggers the release of the poison gas.

G: The problem with Schrodinger's cat is, how do you know whether it's in that box being alive or dead? Has it been poisoned before you open the box and look?

H: Which is akin to Berkeley's debating whether a tree falling in the forest makes any noise if there's no one to hear it fall.

G: And what does this have to do with your one essential problem?

H: The problem with Schrodinger's cat is a problem of things and change. If you open the box you'll see immediately that the cat is either alive or dead. If it is dead, then it has changed. If it's alive, then it's still the same thing that it was. But, what triggers the change if it dies? Whose observation? The cat's? The experimenter's? Maybe that of someone watching the experimenter open the box? Someone watching the watcher? And what state is the cat in while it's waiting for someone to make an observation? Semi-dead? Semi-alive? The questions we're raising with this paradox all revolve around how and why things change, and exactly when one thing, via change, becomes something else. You have, in effect, re-stated the particle/wave problem. And you have also illustrated the difficulty of reconciling universals with particulars. Because while the cat is still inside the box, it's only an ideal cat, like an ideal chair or any ideal form. It doesn't really exist, for the experimenter. But once he opens that box and peers inside, the cat becomes a particular cat, with very specific physical properties which include aliveness—while the ideal cat was neither alive nor dead.

G: I follow that, but I don't see what it has to do with people directly apprehending God's will.

H: Take the Gnostic view, which is that human beings can find God inside themselves, and don't need to have him interpreted. When you're looking inside yourself, searching for God's word, electrochemical impulses are racing up and down your neurons. Whether you're feeling God, or intuiting Him, or thinking Him, or just sensing Him, your brain is active, and you're finding Him *in* the activity. You're finding God in the *way* those electrochemical alterations are occurring. Mental activity, when it comes down to it, is nothing but things changing. So where does God enter into it? There is only one possible answer: He is present in *the way things change*.

G: I'm beginning to get your point. The Gnostic would argue that since there is nothing to reality but things and change, then when he directly apprehends God he is finding Him in the patterns by which change occurs. If God governs change—and it would seem by definition He must—then by understanding the relationship of things and change you'll also understand God's will. And the Gnostic will further claim that this understanding can take place on some deep unconscious level.

H: While the cleric would argue that the information is incomprehensible unless filtered through the articulated word—the gospels. Which, according to them, is where God made His will fully conscious and cognizable.

G: If I may summarize? The most profound questions that haunt humankind may superficially seem distinct from each other. But they aren't. They all derive from a deep curiosity about the nature of reality itself. They are only different ways of asking the same question.

H: Exactly. And you cannot tackle any one of these questions head-on. That is what makes them so profound, so immune to resolution. In fact, you can't tackle them until you have an explanation of the relationship between things and change. It's the brick wall up against which all deeply abstract questions must run. Why do we die? What is the soul? Am I real? Why should I try to behave ethically? We are at a point in history where physics has so precisely documented what's happening in the microcosm and the macrocosm—the atoms and the galaxies—that we have set the physical limits on our universe. We can now try to evaluate it as a whole. And draw theoretical models, of which this is one.

----Three----

Grana: So, then, what is a thing?

Heidrich: A thing is a representation in three dimensions of balanced values in four dimensions.

G: And what is change?

H: Change is modification of the representation as one or more of the four values increases or decreases.

G: And why do things change?

H: Things change because time goes continuously forward. As time goes forward, its dimensional value is increasing. That rising value must be continuously fed into the three-dimensional representation. While there can be local causes of change, they derive from this general cause.

G: And how is time fed into the representation?

H: Through the curved motion of things in their three-dimensional space. Curved motions allow the formation of dimensional balances. How do you achieve stasis without compromising motion?—by curving the motions about each other in such a way that a "system" is formed. For example the solar system, for example the spinning of a gyroscope, for example the interlocked spinnings in all the atoms of your body.

G: And how do such dimensional balances represent increases in the value of time?

H: In the absence of such systemic balances, in a hypothetical state of chaos, there is no way to distinguish time. Passing time is only perceptible with relatively fixed spatial distances. Therefore there isn't any way to show increases in the value of time in the three-dimensional image except by developing the pattern of spatial balances. If the representation is to be an accurate one, it must reveal the amount of passed time by the symbolic forms that it contains. If it does not, it will not be zero-sum.

G: So a system of dimensional balances is a symbol.

H: Yes. And a symbol preserves a record of a certain amount of passed time.

G: And such a symbol is an integral part of the three-dimensional image.

H: Yes. And that image is a symbolic reality. Any symbol inhabits a symbolic reality.

G: Which can be cancelled down to Nothingness.

H: Yes. Another way of looking at that image, that symbolic reality, is as an interference pattern. Like any interference pattern, it appears to be motionless. Even though its composite waves and vibrations *are* motion. Our symbolic reality, as an interference pattern, constantly changes, though, because all four dimensions are increasing; time is going forward and space is expanding.

G: So "things" are pieces of this interference pattern. That's why they appear to be static in shape and form.

H: Yes. Even though if you analyze any of these things, you'll discover that under close scrutiny it's nothing but bits of action.

G: And passing time is continuously transformed into symbols in the interference pattern. The things become more differentiated. Symbolic reality undergoes regular change, and the changes tell time.

H: Yes.

G: Hmm.

H: Of course, that this process applies on a cosmic level of events is difficult to prove. Conventional wisdom is that after the big bang there was a great deal of chaotic energy and no order, no negentropy. The energy flux, it is believed, cooled and settled into bits of form— atoms. With more time atoms became stars, heavier atoms were created in the stars, and we ultimately got what we see out there now when we look at the night sky.

G: The scenario seems consistent with your process.

H: Except that conventional wisdom argues that form is gradually crumbling. In the distant future no form would remain at all. I'm proposing, on the other hand, that at whatever point you slice across time, you'll discover an interference pattern. No matter how far into the future you go, no matter how far back into the past. That interference pattern gives you a three-dimensional diagram of the values in four dimensions. In a universe where spacetime expands and those four values steadily grow, symbolic form, rather than diminishing, ought to evolve.

G: In brief, evolution of form would be eternal.

H: Yes.

G: And that contradicts the conventional wisdom of physics?

H: Well, we're talking about systems here, and the second law of thermodynamics implies that all systems run down eventually. Entropy always increases at the expense of negentropy. Since the visible universe is made up of systems, it will itself eventually run down. Unless there is a lot more mass in the universe than what we perceive, it will expand and cool to a final heat-death. The second law of thermodynamics is considered an inviolable rule of physics. It abides with no exceptions.

G: But you're taking exception to it.

H: Up to a point I agree with it. Physicists are saying that the universe is made up of forms and free energy. The forms disintegrate while the free energy increases. Well, that in itself is inarguable. But I believe there's a corresponding process that cancels out the thermodynamic decay.

G: An evolution of form?

H: Logically, one seems compelled to predict heat-death for the universe. But Thomas Malthus logically predicted overpopulation of the planet. Everyone once agreed with him; now everyone qualifies Malthus' conclusions. The simple fact is that just because something is logical doesn't mean it will happen. More often than not it doesn't, which is why human beings are notoriously unsuccessful at predicting the future. Look at history. Lots of things that logically should have happened never did. It would have been logical to assume that syphilis would wipe out Western Europe a few centuries ago, but it didn't. Why? And why was Malthus wrong? Why hasn't exponential population growth already made this a barren planet?

G: In both instances, because new social factors counteracted the disaster. Syphilitic infection was slowed by the use of condoms, by altered sexual mores, and by antibiotics. As for Malthus' predictions, birth control and abortion have dramatically slowed population growth; in some developed countries, there's actually negative growth. Plus, there's been a revolution in agricultural methods.

H: In other words, the system itself changed. It evolved in response, and it survived.

G: But that merely pays tribute to the ingenuity of human beings.

H: No, it pays tribute to the ingenuity of reality. Look at how bacteria have mutated over the past decades to become immune to our antibiotics. And bacteria don't even have brains.

G: Granted. But on the one hand we've got living beings, which can evolve, genetically or socially or technologically, and on the other we're talking about stars and galaxies and cosmic dust, which can't.

H: But both types are systems, and composed of symbolic forms. They're both subject to the symbolic transformations. And one of the symbolic transformations says that new form harnesses entropic motion.

G: All systems evolve into new form, regardless of whether they're living systems or not?

H: Yes, insofar as they make the necessary observations that lead them into the future. If they don't, they won't survive, but some other new form will take their place, because by definition the future contains more passed-time than the present, and that value of time must be expressed in the apparent form. The future—some future—is going to occur, and the symbolic reality of the universe is going to be different than it is at this moment. If the system undergoes modifications that allow it to remain integral to a zero-sum universe containing more passed-time, then it will perceive itself as having survived. If not, then it won't matter, except maybe to the system itself as it is dying out.

G: Let's distinguish between living and non-living systems here, shall we?

H: Both types make observations. And observation doesn't have to be something you're aware of making. It's more like an effect that the environment has on you. And certainly a non-living system like a star or a solar system is affected by external events. But the non-living system isn't aware of making such observations, at least not in any sense we comprehend. Awareness of observing presupposes some manner of internal modeling of what's going on. Such modeling goes on in DNA, and in the brain, of a living system.

G: Which provides the living system with some control over what observations it makes.

H: Hence, some choice as to whether it makes it into the future as a recognizable descendant of itself. Particularly so if this living system has consciousness. We human beings are conscious. We can choose to make observations that will gradually modify our genetic and cognitive models into the future. There are futures out there in which we would find no discernible continuity of our forms.

G: You are nearly saying that those futures already exist.

H: Well, they do, I suppose. If reality is subjective, as I'm arguing, the only difference between a universe that does exist and a universe that doesn't exist is the point-of-view you're taking. The fact that we have "chosen" this universe is the only reason it appears to be real. Surely there are other universes in which we've already bombed ourselves into oblivion. But any of them can only exist at all insofar as there is some point-of-view within it that makes it real.

G: And if mankind becomes extinct, this universe would vanish?

H: If humankind became extinct, this universe would continue on pretty much as before, because this planet teems with complex life. But let us say for sake of argument that the bombing was so thorough that no life remained on this planet, and that there weren't any alien life-forms out there. Then, theoretically, the universe as we know it might cease to exist. In a zero-sum universe, point-of-view is everything.

G: I find the notion of annihilating the universe along with us even more unnerving than the thought of nuclear war itself. It's absurd.

H: Much of what I'm saying is absurd. Unfortunately for our conventional paradigms, this theoretical description does not contradict the facts as we know them. Our conventional belief is that the universe is mostly chaos, and the few glimpses of form we see are freakish. Accidental, and quite irrelevant to the rest of the universe. The longer I muse on that, the more absurd that cosmology seems. More intellectually satisfying is the idea that point-of-view evolves with observations as time passes, and that without point-of-view there would be no universe at all.

G: Is there significance here to what the new science of fractals tells us?

H: Of course. It informs us that symbolic form can generate chaos. Symbolic form, in this case, specifically means algorithms. But human beings are symbolic form, as well. This suggests that the chaos evident in the universe may be causally generated by us—that is, by our point-of-view. After all, what is chaos? If it can be organized mathematically by fractal equations, then it's not as senseless as we are accustomed to suppose. Chaos, it turns out, is merely pattern that is spread over a wide period of time.

G: Chaos is pattern we can't see from our point-of-view?

H: Oh, certainly. In a given moment it is purely pattern, an abstract mathematical equation. But stretched out over time, it disperses into chaos. It's still a pattern, but not one we readily perceive. The same with any symbolic form; it's visible because it collapses time.

G: Hold on. What does that mean, "collapses time?"

H: Well, collapsing time, or a spatial dimension, reduces it to zero. As with the photograph, in which depth is collapsed. Collapsing allows representation of symbolic form in fewer dimensions. It's not a high-tech process, we all do it all the time. Any static representation collapses time. Most collapse at least one spatial dimension as well.

G: Give me another example?

H: A video of a football game collapses the dimension of depth. You can still read depth into it, but the fact is that the video itself is pre-served on a two-dimensional medium while the game was played in three spatial dimensions. Each frame of the video has only two spatial dimensions, with time collapsed, but when you run the frames sequentially at normal speed, it appears as if time flows. How about those chalkboard drawings we've all seen where an individual play is mapped out, with circles and arrows? That's a two dimensional representation of what's going to occur in four dimensions. Time, as well as one spatial dimension, is collapsed.

G: The reason I asked for clarification is that I recall you used the term "collapse" in discussing the Schrodinger quantum wave functions. The wave function collapsed, and the particle appeared.

H: I don't want to digress too far here, but the word is the same because the concept is identical. When you collapse a quantum wave function, you collapse *all* its dimensions. We spoke of a subatomic particle as being a sort of blur. "Observing" that particle collapses the blur into a dimensionless point. Any observation collapses dimensions. And, of course, a representation is an observation.

G: So, since our frame of reference, our point-of-view, is a representation of four dimensions in three dimensions, then...

H: Then, our symbolic reality is a panorama of collapsed time. And as a three-dimensional panorama, it stands perfectly perpendicular to the flow of time. The trouble with chaos is that the pattern is not perpendicular to the flow of time. Hence it isn't visible. But it's there, nevertheless. All you have to do is modify your point-of-view—your frame of reference, or symbolic reality—to see the pattern.

G: I have trouble grasping that.

H: Let's consider an analogy in fewer dimensions. Visualize a glass box with hundreds of dots suspended in it. When you peer into the box from the sides, you see no order whatsoever. No pattern. The dots seem to be randomly scattered. Okay, let's say we're working with two spatial dimensions plus time, instead of three plus time.

We'll have the width and length of the box be the two spatial dimensions, and the height of the box will be the dimension of time. So the height will be a certain period of time, let's say a year. The box is one year high. The random dots, then, are scattered through both space *and* time. The dots represent events occurring during the course of a year, within two spatial coordinates. No pattern. However, it is possible to collapse the year's duration down to one moment. And that process, of course, is what we use when we make a three-dimensional reality out of four dimensions. When you perform such a transformation, you're establishing a frame of reference. A frame of reference is flat—perpendicularly flat—against the flow of time. Okay?

G: If you mean that what we're trying to do is bring the chaos into a frame of reference so that it displays a pattern, okay. How do we do it?

H: You simply lean over the box and look down into it. Lo and behold!—you see a cartoon picture, just like the Sunday comics used to be, made up of little dots! So those dots weren't randomly placed. They were random only so far as your first frame of reference was concerned. In that one, the dots were spread out over a duration of time.

G: Ah. So, to relate this to fractals, the chaos is the side view of the box, while the equation is the cartoon picture. When you look down into the box, you're seeing events over a year's time compressed into one moment. Human beings looking at the universe are seeing events spread over ten or twenty billion years compressed into a moment. As new events keep occurring, they must be immediately integrated into this image.

H: But the integration is tricky, because the image must continue to be a zero-sum equation, and it must remain perpendicular to the flow of time. If you do it right, you perceive a continuity to your symbolic form.

G: And, as far as things and change go, things are the pattern involved in the point-of-view, and change is shifting in the patterns perceived?

H: Yes.

G: Then everywhere—and everywhen—there is total chaos, as well as endless patterning. It's all in the point-of-view with which you perceive.

H: Exactly. Any point-of-view will perceive a great number of patterns, but also a great amount of chaos. As the symbolic reality evolves, some of the patterns will move into chaos, which is entropy, and some of the chaos will focus into patterns, which is negentropy. There's a constant shifting of perspective going on, and it's necessitated by the flow of time. Passed-time must be incorporated into whatever patterns are perceived. Since perception is in point-of-view, the point-of-view itself is what is really changing. It is evolving. The patterns are made up of *things*, and the shifting of perspective is *change*, and the dynamics of the process are reducible to a single set of symbolic transformations.

----Four----

Grana: We've been discussing how things change. But there are things, and then there are *things*. Our world is comprised of ideas and facts and opinions, with attitudes, personalities, relationships, and so on, besides the physical realities.

Heidrich: And, these intangible things behave the same way as the tangible things—according to the same processes of transformation.

G: Why?

H: Because they are elements in symbolic realities, albeit intangible ones. What is a symbolic reality?—it is a coherent and integrated representation of the zero-sum. Whether it is the fundamental representation, made of atoms and molecules, or merely a description of it, or a description of the description, is irrelevant. All such models are constructed by the symbolic process—that is, by collapsing time at various points to make images in a spatial matrix. Our universe is such a model. But so is a theoretical universe in the computer of a physicist. There is a common blueprint to all such models, to all symbolic realities.

G: How, then, do you distinguish one symbolic reality from another, in a given situation?

H: By the point-of-view. After all, that's what a symbolic reality is—a point-of-view. We have tentatively established that objective reality doesn't exist, correct? Then, whatever reality is perceived is subjective, and is in fact determined by the nature of the perceiver. Even if you're referring to what we call the physical universe. Twentieth century physics reverberates with the assertion that physical reality depends on point-of-view. In quantum physics, subatomic activity depends on how it is measured; in relativity theory, the cosmos is a warped space-time continuum. So, in both paradigms reality is inseparable from the perception of reality. Unirealism is a broadening of the same general theme. Our world of *things* and *change* can be fully described and explained in terms of subjective perception. In fact, that is the only coherent way to describe and explain it. And what I call symbolic reality is the conceptual linchpin.

G: Then let's talk more about it, shall we? Starting at the beginning-- the physical universe is a symbolic reality, correct? It is a symbolic representation of true reality.

H: Yes. True reality is, in fact, Nothingness. Objectively, it doesn't exist. So, the physical universe is a symbolic reality that represents the zero-sum Nothingness.

G: And any descriptions of that physical reality, they are symbolic realities as well? For instance, after forty years of living I have put together a mental model of the world out there. It's there in my mind every waking minute of the day. With it, I negotiate my way around the real world, deciding what route to take to the office, when to buy my wife flowers, and so on.

H: That's your personal symbolic reality. Your point-of-view and frame of reference. It's a miniature model of the world, a sort of virtual reality.

G: Okay, so now we have two symbolic realities. One is the physical universe. The other is my map of it. Why should the former be tangible and the latter intangible?

H: Now that is a fascinating question, isn't it? According to the principles of unirealism, all symbolic realities evolve and sustain themselves and change by energy transformations. So why should one possess tangible energy and all others intangible energies? In unirealism, the term *tangible* refers to the condition of being logically prior. Our physical world is the primary model of the zero-sum, and your personal symbolic reality is a secondary model. Since the primary is logically prior to any subsequent abstract model, it has a stubbornness of form. You can make a chair disappear in your mind, but not the chair you're sitting on, because its apparent existence is a prerequisite to your perception of it. By religious doctrine, of course, God can make that chair vanish, if He so wishes, but that's because religious doctrine considers Him logically prior to the physical universe. After all, He created it. Please forgive me if it seems a digression, but I would emphasize here the functional complexity of our physical reality. At the risk of stepping on God's toes, it isn't unlikely that if that chair were to vanish—acausally—the entire universe would crumble. The chair, of course, could causally vanish, if, let us say, Scotty beamed it up, but in such event no laws of physics are violated. The form of the universe is not accidental, and not only can you not randomly vanish things, you can't randomly tamper with the overall design. We don't yet know why the value C is 186,000 miles per second, or why Planck's constant has the value it does, or why the four forces of nature exhibit the relative strengths they do, but almost certainly they are exactly what they must be for our universe to exist. And such factors are interconnected. As with the ecosystem, altering any one factor causes a cascade of revisions throughout the unified whole. Physicists are trying to understand how this whole jigsaw puzzle fits together, but there's so very much they don't know.

G: So tangible is tangible because it can't be anything else.

H: And nothing unnecessary exists. Whatever has to have happened for us to be here has in fact happened. The evidence of those events constitutes our physical reality. That may seem obvious, but it has a hidden implication, which is this: anything which did not need to have happened *has not happened*. In short, the past is unfinished.

G: Surely you're joking. Past *means* finished.

H: Not at all. We know for a fact that at least at the quantum level, some past events are not yet decided. They still may turn out a certain way, but they haven't yet. Which, parenthetically, explains why we can have free will. Free will with which we can alter physical reality, if we so choose, provided we don't violate causality. Free will with which we can alter our abstract models of reality, which are far more forgiving of careless construction.

G: More forgiving because they are intangible.

H: Yes, and prone to error, distortion, inaccuracy. Whereas physical reality can contain no errors. It must be at every moment unfailingly accurate, as a depiction of the zero-sum. You routinely correct errors in your personal symbolic reality, for instance, in the process of maturation and learning. Such a model is imprecise, either due to misinformation or lack of information.

G: And it is up to me to monitor it for mistakes.

H: Yes, and why? Because it makes you effective. You've been constructing it since infancy. It is the sum of all your learning about the world, starting with how to walk, to talk. How to play cricket and charm an audience, how to be a journalist and a father.

G: So the development of a model is purposeful.

H: A model can be defined as a representation similar in appearance to the original. For example a fashion model, who is supposed to demonstrate what you'll look like when wearing an outfit. But it isn't merely superficial appearance that's important. After all, you can build mathematical models, too. So, when a model is purposeful, the purpose is to facilitate predicting behavior in the original. If you think about it, a model is constructed out of symbols that represent aspects and behaviors in the original. In the case of the fashion model, the exaggerations of her physical features make her ideal for conveying symbolic meaning. The high cheekbones, sleek hair, full lips, and so on. And when the female model glides down the runway, or when a male model stands brooding on a moonlit pier, what the designer and advertiser are suggesting to you is an impression of how you'll look during similar behaviors. The model is modeling you, and making it quite a flattering image.

G: A model, then, is composed of symbols. Even the symbolic reality that is the universe—that is composed of symbols?

H: Atoms, molecules, galaxies—these are symbols, yes. As are words, images, numbers, archetypes, and other constructs. Human beings assemble symbols into models purposefully, and employ them purposefully. But reality generates models, even when there's no conscious purpose or obvious intent.

G: How does that happen?

H: Through an accumulation of observations.

G: Observations of what?

H: Of the environment. By environment I mean a logically-prior symbolic reality. Observations of activity in that environment accumulate and integrate to form symbols, which trigger new observations, and eventually a perceptible symbolic reality emerges. If there is sufficient energy available in the original, secondary symbolic realities will bloom as readily as flowers.

G: Such as the appearance of life-forms in a primordial soup.

H: Oh, yes. The ecosystem, which some call Gaia, is a secondary model of our universe. You yourself are a miniature symbolic reality that has blossomed out of the ecosystem. Which means that you are also a model of the universe. All these models overlap to some extent, with hierarchical degrees of complexity and somewhat different emphases.

G: It's peculiar to imagine myself a model of the entire cosmos.

H: But you are. If a being from an alien universe captured you and transported you home to his alien laboratories, scientists there could deconstruct you, analyzing your physique and atomic and cellular make-up, and from the results determine just about everything there is

153

to know about our universe. The number of dimensions, the relative strength of the fundamental forces, the values of pi and h, the age of our universe—all that. They could infer the existence of galaxies, quasars, and black holes. They could speculate accurately about what other species inhabited our planet. They'd figure out everything that was significant in the twenty billion years history of you, because there's nothing in your make-up that wasn't born out of observations.

G: A minute ago you said the emergence of any symbolic reality depended on sufficient available energy. Does this mean that energy itself can be defined in terms of symbolic transformations?

H: Yes.

G: Okay. Now, say that we have a symbolic reality, and we're going to generate a model symbolic reality from it—a derivative model. You said that can happen if the original has sufficient energy in it.

H: You take potential energy from the original, or logically-prior, symbolic reality, break it loose so that it becomes kinetic, and then rearrange it into symbols for your new symbolic reality.

G: Let's follow this process in a particular case, the emergence of our ecosystem, with us in it.

H: The sun is our primary source of energy. We draw some energy from the internal heat of the earth, and from cosmic rays and such, but it is really the flow of solar energy that drives the ecosystem. That energy travels as electromagnetic waves, and arrives here as photons. Our ecosystem, as a model symbolic reality, is evolving through observations. And every one of those photons that strikes the ecosystem is an observation. When you gaze at the sun, you're absorbing photons from it. You're observing. And, in making all those tiny observations, you're converting the sun's potential energy into kinetic energy, which you then incorporate into yourself, making it potential energy again. If the sun weren't there to provide a source of observations, the ecosystem would shrivel and die.

G: It sounds almost as if we are causing the sun to burn by looking at it, by feeling its warmth.

H: Well, would your food burn if you didn't actively digest it? In digesting it, you're converting its potential energy into kinetic energy, which the body then deploys to sustain itself against its own entropic decay. I think there's a reasonable argument that the sun wouldn't burn if there were nothing else in the universe to observe it, thus causing it to burn. Just like Berkeley's contention that no one hears the tree falling in the forest.

G: We have talked about three symbolic realities now: the physical universe, the ecosystem, and my personal cognitive model. How about some others?

H: A functional society has a collective symbolic reality. Any good novel creates a symbolic reality. A weltanschuang, or world-view, can be classified as a symbolic reality. Any specialized discipline—biology, physics, economics—that models larger reality. In fact, any integrated system of symbols that arises naturally from a prior integrated system, and attempts to model it, is a symbolic reality.

G: And each one borrows energy from the original, via observations.

H: Presupposing that the original has sufficient energy to energize the model. For instance, the novelist invests an enormous amount of energy in his work. The novel is built up out of observations, which are turned into symbolic form, which forms are then arranged into an abstract model of transactional reality. When a novel is finished, of course, it doesn't require a continuous input of energy to sustain it. Generally, though, a symbolic reality is an ongoing thing, and its energy provider must be ongoing as well. Like our ecosystem. Like our government, which abstractly models our society, and which requires the energy inputs of voting, taxes, and data.

G: A community would be a symbolic reality, even a neighborhood. A university, a corporation, even. A culture.

H: And a counterculture. It's interesting to note that the Sixties' counterculture was sustained when the larger society had enormous amounts of available energy. A counterculture is a societal luxury.

G: A family?

H: A family, a tribe, a civilization, even a species.

G: Symbolic reality is a very broad term!

H: As it must be. We're using perhaps a dozen concepts to explain all of reality and how changes in it occur. We're reducing reality to a relationship between nothing and everything. Unirealism says, in effect, that point-of-view is the *sine qua non* of reality. Without it reality is a nonsense term. But reality has a great many meanings for us. Physical reality, interpersonal reality, the reality of the business world and that of politics, the social realities, the realities of the environment, and on and on. The definitions and transformations of unirealism embrace all those meanings of the word. We want to understand the dynamics of reality, and the only way to apply a single set of dynamics to all those varieties of reality is to generalize a concept of symbolic reality. That is, a model based on a point-of-view.

G: And all symbolic realities are to some extent models of each other, leading back to a zero-sum Nothingness. *Things* and *change* of any kind follow the same transformations everywhere. It's just a matter of designating which symbolic reality you assign them to.

H: Yes. All disciplines of human knowledge, from anthropology to zoology, are inquiries into what things are and how they change. The tragedy of contemporary academia is not specialization in and of itself, but rather the failure to perceive a common theme. When the Greek philosophers came to an intellectual dead-end it wasn't just because they had limited themselves to deductive reasoning—neglecting empirical data—but also because they didn't specialize. Over the subsequent centuries, science developed the implicit belief that the big picture was unimportant, and that specific results were what counted. The marvels of microbiology, genetic engineering, medicine, quantum physics, and all our technologies—such is the bounty of that daring attitude.

G: You don't hope for a return to a time without specialized fields. You only outline the umbrella under which all such disciplines stand.

H: I do this necessarily. Economics, linguistics, history, psychology—they are in their end-times because they have gone about as far as they can go. Each has its own jargon, but the concepts and terms are orphans. Untethered, they hang there in a void. However, once you fit them into the scheme of unirealism, they have a foundation. Once you acknowledge that the universe is zero-sum, then and only then does the concept of *ego* have any discernible relation to the concept of *quasar* or the concept of *virus*. You can not only trace the significance of any one concept back to the initial zero-sum from which all reality has emerged, you can lock each concept into the context of symbolic transformations. Given this foundation, every idea known to humankind can be explained in a single, coherent, system of meaning. And the beauty of the system is that it is drawn out of nothing, and needs no original cause.

G: Everything is symbolic, and the symbols can be traced back, through various energy conversions, to the very beginning of time?

H: Exactly. We conjecture in symbols, and we exist as symbols, but symbols are only abstract representations emerging, ultimately, from Nothingness. Think of the disciplines. What is linguistics?—the study of cognitive symbols. What is philosophy?—the study of how conceptual symbols relate to physical reality. What is economics?— the study of symbols of exchange. What is mathematics? What is art? What is music? All such fields are studies of symbols, and all the technologies are applied techniques of using symbols. The essential component of reality is the symbol. If you want to analyze reality, you analyze what symbols are, and once you understand the nature of the symbol, you understand the structure of all knowledge.

----Five----

Grana: Reality—of any kind—is composed of symbols?

Heidrich: Yes. Any symbolic reality is an interference pattern, and the symbols are like standing waves. They're three-dimensional constructs in a four-dimensional medium. They're motionless because they've had the forward motion of time extracted from them. They just sit still.

G: Which is what makes them, for us, real.

H: That's fair to say. A symbolic reality, at any moment in time, is a motionless image. Motionlessness is a crucial part of the definition.

G: Now, a symbol is real...

H: Only in terms of the point-of-view. It may not be real from another point-of-view. But as long as that qualification is understood, yes, you can call a symbol "real."

G: Okay. A symbol is real, and atoms are symbols. But I seem to recall you saying that the subatomic particles are not real.

H: Subatomic particles are not real, by my definition of the word. You recall that we talked about the subatomic particle as something of a blur?—an uncertainty? Well, if you overlap some of these blurs properly, you can force a particle into focus, and out of its uncertainty. You can collapse its quantum wave function—the qwf, which states its various probabilities of location—and the particle enters symbolic reality. For that briefest moment the particle stands still. What you've done is created a node in the interference pattern which is symbolic reality. It can be argued that at that point of qwf collapse the particle is real. But, considering the fact that it instantly vanishes again into a blur, while the illusion of reality remains, you must conclude that the particle is not integral to the illusion. It suffices for it to be no more than a blur of uncertainty.

G: A qwf collapse is an observation?

H: Because when you collapse a qwf, you stop time at that location. Normally, subatomic particles are in motion. When you observe one, you freeze it momentarily in time.

G: But it won't stay frozen, because it isn't locked into a dimensional balance.

H: Essentially that statement is true. And if you are able to arrange a few of these subatomic blurs in a dimensionally balanced complex, then you'll have a symbol—an atom.

G: At which point it becomes real.

H: And a node in the interference pattern. Reality doesn't manifest until you're representing in three spatial dimensions the values in four dimensions. So there's no point in talking about anything being real until you've got evidence of the symbolizing process at work.

G: That's with physically real symbols, like atoms and molecules. How about in abstract symbolic realities? Those are also composed of symbols?

H: Yes. Symbols which are formed by the same technique, which is extracting time and then containing it in the representation.

G: That representation is a dimensionally balanced complex?

H: Yes.

G: Which is an integral part of a symbolic reality.

H: Yes. There's all kinds of abstract symbols: linguistic symbols, pictographic, heiroglyphic, iconic, vocal, mathematic, archetypal, musical. There are also signs and gestures and facial expressions. We use all these types of abstract symbols to communicate. All of them are representational. An abstract symbol represents an aspect, or property, or quality of the environment, or an object, so that when you encounter the symbol the effect is similar to encountering the original stimulus.

G: As an example, when one hears the word "tiger" one gets a sense of the original—the real tiger. When one sees a wedding ring, one gets a sense of the wedding ceremony and what it signifies.

H: Which is to say that one re-experiences the represented gestalt. Usually this re-experiencing happens subconsciously. And one doesn't always experience every aspect or element of the original ge-stalt. Re-experiencing the meaning of the symbol is like replaying it on an internal screen. Whether it is a visual image, or tactile sensa-tion, or emotional response, or auditory impression depends on the person and the circumstances. In any event, at some deep level there is a sensory experience of what the symbol stands for. Even with highly abstracted symbols that admit of no concrete original this is true. And conscious thought, as well as unconscious thought, flows on the current of such symbols.

G: That would suggest that intelligence and facility with symbols can be equated.

H: Well, standard intelligence tests measure one's facility with sym-bols.

G: Can we also surmise that it's the ability to manipulate abstract symbols that sets Homo sapiens apart from other species?

H: It might be desirable to break that distinction down into two fac-tors, which are related. One is that human beings have extremely sensitive nervous systems, and the other is that human beings have a talent for symbolizing the information in that veritable barrage of sensation. Probably these factors evolved simultaneously. If a creature had an overly sensitive nervous system but not comparable ability to symbolize in response to it, the thing would likely go mad. You're familiar with *The Miracle Worker*, about Helen Keller, of course. As for the ability to symbolize, that involves not merely brainpower, but also dexterity, facial expressiveness, and sophisticated vocal organs. Without those physiological features, you can't invent, manipulate, and communicate symbols.

G: Apparently dolphins and chimps have been taught how to use symbols.

H: They've been taught how to use our symbols. But all species employ symbolic behaviors in their natural habitats, since they are all sensitive to their environs.

G: Is consciousness unique to people, though?

H: There is a *type* of consciousness that is unique to our species, you could say that safely. But debating degrees of consciousness and when or where it's being exhibited is tricky, even when you carefully define your terms. For example, DNA itself is a genetic code, highly symbolic; hence all life-forms use symbols. You may say, well, that's not really consciousness, is it? But isn't evolution a process of observation and re-coding? I'd call that thinking on a genetic level. Notice how microbes adapt and become immune to our poisons—that's what antibiotics are to them. Our own species observes that some substance is a poison, and learns to avoid it, or how to neutralize it. So does the microbe, by evolving. And they not only mutate to evolve, they even borrow genetic material from other microbes, building a library of defenses.

G: But learning alone isn't consciousness. Consciousness uses logical reasoning to solve problems. Reasoning that can be communicated.

H: If you're reluctant to acknowledge that bacteria and viruses are conscious, consider the consciousness demonstrated by honeybees, and ants. They reason very logically, although they do it on a collective level. And they communicate with each other quite effectively. So what is different about human consciousness? It's true that we are conscious of ourselves in a way other creatures don't seem to be. We have self-symbols, and use them to think about what and who we are. That may be unique, or we may discover that whale songs are meditations on identity.

G: I suppose it doesn't matter, really. I'd just like to believe that abstract symbols are the invention of human beings.

H: But they aren't. Look at the peacock. The peacock's tail is symbolic. Coloration, markings, and other extremes of appearance are symbolic, in any species. Instinctive displays, mating rituals, and aggressive poses are all symbolic. The roar of the lion and the chest-beating of the gorilla are symbolic. When we say such behaviors are instinctive we're saying that they aren't consciously exhibited. The single peacock when it ruffs up its tail is not consciously using it as a symbol. But the peacock species is. That array of symbols has arisen out of observations made by the species. In fact, if you look at life-forms through the subtlety of symbolic theory, you notice that all the parts of any organism are symbolic. What is the human hand, but a representation of ten million years of environmental conditions and circumstances? Ten million years of observing go into the architecture of that hand, perfecting it to its most efficient shape. The same with internal organs. Any single organism, any life-form, is a collection of symbolic responses to what happened to earlier organisms in the species. Evolution is a series of symbolic responses to various generalized gestalts. So a species is a kind of conscious thought, with genes as symbols. The species thinks very slowly. But it thinks. Similarly with the ecosystem as a whole. Calling it Gaia imputes an abstract consciousness to it.

G: One might hypothesize, then, that a symbolic reality that actively self-replicates, by developing symbolic responses, is itself conscious?

H: That's fair enough.

G: Okay, how do such symbols get developed?

H: In the same way that atoms develop. By taking the flow in four dimensions and representing it in three spatial dimensions. In other words, by extracting time and containing it in the symbol.

G: But how does an abstract symbol contain time?

H: Imagine that your life is a movie film projected at a theater. Allow that there are certain recurrent scenes in this film—scenes of you eating, for example. As long as the film is rolling, time is going by. Okay, you stop the film during each of the recurrent scenes—similar gestalts. You clip one frame from each scene, set them aside. Now you generalize them: you pile them atop one another, shine a projection beam through them, and get a fuzzy image on the screen. You focus it. Now you have a generalized gestalt, and as soon as you name it you have a manipulable symbol.

G: We'll name it "eating."

H: Any kind of physical behavior can become a name. Human beings usually vocalize during both work and play. It's very natural for us; we chatter and babble all the time in a natural setting. But you could also name a behavior with a hand gesture or a visual symbol. Or all three. Evolution names such stop-action generalized gestalts with genes. It then continues using them symbolically, just as we do our learned articulations. Bear in mind that this process of inventing symbols from observations can only go as far and as fast as the sensitivity of the nervous system allows. After all, an image may be very fuzzy. Focusing it is seldom easy, so the acquisition of new symbols doesn't happen overnight.

G: How, though, does this symbol contain time?

H: It's plain enough that a given genetic structure contains history, but let's look at the symbol of you eating. By freeze-framing the film at a particular moment you have removed time from the four-dimensional equation that describes the unfolding events. You have performed surgery on the flow, cutting out the temporal dimension and inserting it into the three-dimensional shape. Basically what you've done is taken a chunk of the four-dimensional flow and made an image of it from which the dimension of time can be inferred. In much the same fashion that you can infer depth in a two-dimensional photograph. I'll give you another example. Do you play cards? Bridge, pinouchle, hearts?

G: All of the above.

H: At the end of a hand, you have little stacks of playing cards on the table, each of them representing a trick. Each stack is a symbol, a representation of that short period of time during which each player laid a card face-up on the table. In theory, you can reconstruct the last game by analyzing the evidence in those tricks. So the symbols contain the history. The symbols contain time. That's what symbols do.

G: And if you didn't make symbols you wouldn't be able to hold onto the past. You'd have no record of events. You'd be like the brain-damaged patient who can't recall anything that happened more than fifteen minutes ago, who lives entirely in the present.

H: Sure. But generating symbols is what reality does. Even primitive cultures have oral traditions, and kinship trees that go back many generations. Recording the present in symbols is the most natural process in the world. It is not inaccurate at all to say that it is the *only* process in the world.

G: Hypothetically, if there were no such process...

H: Then who would ever know? No one would exist, because we are all made up of symbols. No universe would exist, because it is made up of symbols. If this process didn't take place, the present wouldn't have any past. There would be no symbolic reality at all. All of time would be jumbled up together. The past doesn't exist, except insofar as the present symbolic reality says it does. And it expresses that past by the symbols it contains.

G: In other words, symbolic reality is the record that history occurred.

H: And the time that has passed is bound up in symbols, which are the building blocks of any symbolic reality.

G: So just as you can look at the genome of a human, and see the history of humanity coded there, so also you can look at the night sky and see the history of the universe coded there.

H: The stars, the clusters and galaxies—they are the written history of our universe, yes.

G: I do believe I'm beginning to understand what unirealism is. Here, let me try to articulate it. The world is made up of symbols and symbolic form. These symbols do not have inherent realness; they are real in a context. The context is symbolic reality. Symbolic reality is a point-of-view description, and may also be referred to as a frame of reference. Its building blocks may be tangible or abstract. The important thing is that overall, it is a map of four-dimensional reality. As for maps like this, they may use different symbols, different scales, and they may be either elaborate or crude, they may be individual or collective. Any map, however, is a map that has evolved from the primary map. The primary map is the physical universe itself. And, as time goes by that primary map changes. It makes observations and evolves in symbolic form. So do the secondary maps, as long as their source of energy isn't diminished. So far, so good?

H: So far, very good.

G: Expressed another way, the world we experience is all composed of symbols. Atoms, molecules, genes, concepts and ideas, names, solar systems, life-forms, and whatever else you discover—all of it is just a great agglomeration of symbols. It may seem to the untrained eye to be an unorganized agglomeration of symbols. But if you look

165

at such symbols as being integral components in various symbolic re-alities, you perceive that they're not randomly thrown together. Each symbol is part of a map, an integral part of a map. Furthermore, each and every one of these symbols is created with the same technique. The technique is one of expressing in three dimensions the motions in four dimensions. Why any given symbol develops only makes sense when you consider it as a development in its symbolic reality.

H: You're on a roll, now.

G: Because a symbolic reality is an interference pattern of four ex-panding dimensions, the interference pattern cannot be static. The expansion of the spatial dimensions and the forward motion of time serve to compel a progressive evolution of the pattern. The pattern at any given moment describes, in an infinitely complex equation, the values of the universe's dimensions. The pattern is what gives us form, and as time passes the form changes, by means of observations. The observations are arranged in dimensional balances, which are symbols. That is how you get symbols.

H: Yes, excellent.

G: So, not only are all things in this world symbolic, but all changes in this world involve transformations of symbols. That's all there is, because the world is nothing but a symbolic description that is engi-neered in such a fashion that it can't stop changing. New symbols be-ing created, old symbols being modified or perhaps annihilated, syntheses of symbols being made. All such changes—when inter-preted in the context of the appropriate symbolic reality—occur by rules of a single set of symbolic transformations. And this set of transformations is derived from the original blueprint for making real-ity out of Nothingness. And that's as far as my notes go.

H: It's a remarkably good synopsis.

G: My next question, however, is…

----Six----

Grana: So what? What does this have to do with the rat-race that I'm in? What does it have to do with war, injustice, and the hole in the ozone? Does it improve our children's education to teach them that they're each a symbolic reality and the universe is really nothing? Do the principles of unirealism enable us to live more happily, more productively, more ethically? Or is this just an exercise in philosophy?

Heidrich: If we just go a little bit further, I believe your question will be answered.

G: Okay.

H: Symbolic reality is a frame of reference. It's a map on a three-dimensional matrix and seems to stand still, even as the night sky appears to be dotted with fixed stars. If we focus our attention on any single element in this frame of reference we notice movement, but on the whole it seems to us to be a static background against which things occasionally move. This frame of reference is a three-dimensional cross-section of four-dimensional reality. As with any cross-section, you're eliminating one dimension from the original from which you take the cross-section. If you cut across a tree trunk to expose the rings, you're taking a two-dimensional cross-section of a three-dimensional object, and the dimension you're eliminating is height. Similarly when you set up the frame of reference of a symbolic reality. It's a cross-section at right angles to the original. The cross-section is cut perpendicular to the flow of time, and thus eliminates the dimension of time.

G: As with the tree-trunk, where a cut perpendicular to height eliminates the dimension of height.

H: Which leaves you with a two-dimensional plane. Whereas the cross-section of symbolic reality is a *three*-dimensional plane, perpendicular to the flow of time. The concept of a three-dimensional plane isn't that hard to grasp, is it?

G: Not really. I can think of it like using the pause button on my VCR. The flow of the movie freezes. The dimension of time is eliminated. Leaving what appears to be a three-dimensional tableau—a plane, if you will. A plane perpendicular to the flow of time. That freeze-frame is a cross-section of the movie. A symbolic reality, or frame of reference, is also a cross-section to the flow of events.

H: Changes will occur in the frame of reference as time passes, but at any given instant the cross-section remains perpendicular. It must always be flat against the flow of time. If there is distortion, then the frame of reference is no longer an accurate representation of the four-dimensional reality.

G: Okay.

H: In this frame of reference we call physical reality, any change instantaneously leads to modification of the frame that balances out the distortion. Modification that again flattens the frame. That is why physical reality is so predictable, why we can determine the laws of cause and effect. The frame must, by definition, remain flat at all times. Not for a single instant is the frame of reference *not* flat. The cartoon roadrunner races over the edge of a cliff and hangs suspended there for a few moments, defying gravity, until he realizes what's happened. Then he falls. But in real life he'd be falling the instant the ground no longer supported him.

G: Even when he's in freefall, the frame is always flat? It doesn't just return to flatness after he crashes into the bottom of the canyon?

H: No, it's always flat. The temporary loss of balance in the frame, during roadrunner's fall, is accounted for by his momentum relative to the frame. During his fall, the frame continuously re-balances itself through tiny relativistic effects, including a slight increase in his mass and a slight shortening of his body in a downward direction. Those small physical changes are strictly from the point-of-view of the earthly frame of reference, you realize. That is, from the point-of-view of the viewer, who is standing still. The roadrunner himself does not perceive himself getting shorter or heavier, and even if he were equipped with a portable yardstick and scale he couldn't measure any differences.

G: Einstein's relativity tells us that, right?

H: Yes. Will you forgive me a short digression here? Suppose that this roadrunner of ours ran off the cliff so fast that he attained escape velocity and flew off into space, leaving the earth behind. Out in empty space he'd continue flying eternally, unless he actually flew into something, or close enough to a heavenly body to get pulled in by its gravity. Suppose, though, that he has an uneventful trip. His trajectory won't be perfectly straight, because gradually it will curve with the overall gravity of the universe. Where will he end up, after eons of time? Why, right back where he started. He'll have travelled in a great circle, and if nothing interferes he'll keep going round and round—performing cycles. So even though the roadrunner set out in straight motion, which appeared to become an aimless wandering through the cosmos, bent a bit this way, then a bit that way, by various gravitational fields, he was in fact starting into a cycle. And this is the case with all motion in a frame of reference. The instant you break some balanced energy loose, the frame is already contriving how to make it a balanced part again. In other words, how to convert that errant energy into a cycle, or wave, or oscillation—which are balances. Granted, the nascent cycle of the roadrunner isn't likely to be completed. He isn't even likely to become a moon cyclically revolving around the earth. What happens instead is that he fails to achieve escape velocity, and crashes into the canyon floor. Some of his kinetic energy is absorbed into the atoms of the ground—which, you recall, are themselves only balances of cycles—and some radiates off as heat and light. But even the photons radiating off, if not intercepted, will make great circles in the universe, and thus make cycles.

G: The point seems to be that everything happening in our universe tends to be cyclic, even though we can't always see the cycles.

H: Precisely. Our frame of reference is composed of balances and imbalances—energy vectors that are trapped in symbols, hence potential, and energy vectors that are loose, hence kinetic. But there isn't really a substantive difference between the two types except from the point-of-view, which sees some cycles and doesn't see others. It's analogous to the situation we have with fractal equations.

G: Change your point-of-view, and you can see the invisible cycles. But, if you do, you lose sight of some of the first ones.

H: And, to return to the idea of the frame remaining always flat, the flatness includes all extant dimensional values—or, energy vectors— in the universe, no matter whether they're bound up in balances or apparently kinetic. The frame of reference appears static, because what we generally perceive is the stability of its cycles. But it isn't fixed, it's forever in frenetic change. And that change is noticed in the kinetic motions within the frame. The apparently unbalanced energy vectors. And where these vectors are, at a given time, is part of the flatness.

G: In effect, whenever you change anything in physical reality, all you're doing is rearranging nodes in the interference pattern, but no matter how you do that the frame, or pattern, will remain flat against the flow of time. Even during change the frame is never distorted from flatness.

H: However, abstracted models of the zero-sum equation *can* suffer distortions. And this distinction arrives us at the answer to your question of "So what?" Because while it's true that physical reality must ever remain flat, our derivative models—our abstract symbolic realities—only fitfully respond to the necessity of remaining flat. These models are accurate depictions only insofar as they are flat. Insofar as they are distorted, they are not accurate models. And it is the inaccuracy of our models that causes us all our problems.

G: Wait. Abstract symbolic realities are grounded in the physical reality they imitate, aren't they? They shouldn't be able to distort.

H: Let's look at our tree-trunk cross-section analogy again. Suppose that the crosswise cut was imperfect. The surface isn't exactly planar, but has some ridges in it. Such imperfections are metaphors for distortions in the frame of reference. Now, despite these distortions, if you lean over the surface and look down on it, you'll see an apparently flat cross-section. The distortions, or ridges, will disappear from view. In physical reality such distortions can't exist at all. They're impermissible by definition. But conceptual frames, these derivative models we use all the time in our thinking, they can be distorted, and as long as you insist on looking down at them they still appear flat.

G: Perhaps an illustrative case?

H: I could provide a thousand. In economics, politics, psychology...

G: Economics.

H: Okay. Until very recently, the economic system in this country assumed we could perpetually produce garbage while depleting natural resources. Today, it is obvious to us that we must convert back to natural resource as much of the garbage as is feasible. We understand that now, and have realized that all along we'd been working with a distorted economic model. Of course, the reality was that somewhere out there the garbage was accumulating while the resources were dwindling. So physical reality wasn't distorted. Only the abstract model was distorted. It wasn't flat.

G: Ah, and the distortion was not corrected immediately because people kept saying that we would solve the problem when it got bad enough, right?

H: Sure. That's an instance of finding flatness by looking down on the distortion from above. There are other methods for artificial flattening. Let's consider an example in psychology. Imagine a person who has suffered some horrible trauma and never been able to integrate it. Freud would say that this person has repressed memory of the trauma. Freud would suggest that, as a consequence, the person's libido is blocked. For our purposes, libido is synonomous with energy. Now, in order to function, the person is obliged to perform some adaptive behavior. Maybe, like Lady Macbeth, he washes his hands a hundred times a day. Or, he routinely goes on alcoholic binges, dries out, and swears he'll never drink again. Or, he counts his steps, making sure they always come out even numbers. Or any other obsessive-compulsive behavior that is crucial to sustaining his psychological model of himself. If you look objectively at what he's doing, which isn't easy because such behaviors are often concealed, you perceive that this distorted mental model of himself is being flattened by such behaviors. In other words, he feels the presence of the distortions and performs the behavior to get the model back into flatness, at which point he feels relief.

G: And just exactly how does the behavior return him to flatness?

H: You must understand that such adaptive behaviors are always cyclic maneuvers. As are all ritualistic behaviors. Any behavior like this is composed of a sequence of steps that bring the behavior to culmination. Not only do these behaviors occur with a certain rhythm in a person's life, but each incident by itself has a specific rhythm. It's easy to see this with the obsessive-compulsive, the "one-two-three-four, one-two-three-four" routine that he injects into his life to keep it apparently flat, or the five minutes every hour spent washing the hands. Little rhythms one adds to one's life. Why? For the elementary reason that without such added rhythms the person isn't flat against the flow of time. The interference pattern is out of whack, and this rhythm must be added to it to balance it properly. It's just another technique for artificial flatness.

G: And he can feel that his interference pattern is out of whack, on some unconscious level?

H: Absolutely. It's a truth of human nature that flatness and distortions are instantly recognized, and that while we try to rid ourselves of distortion we also try to prolong our periods of flatness. Were it not for this sensitivity, we'd never advance human knowledge, we'd have no concept of truth, we'd have no morality. Knowledge, truth, and morality all depend upon this inferred standard of reference, which arises from the tendency toward flatness in any symbolic reality. On the other hand, it is a bleak but honest fact that most of the cruelties and injustices inflicted on our fellows are attempts to bring into flatness a particular symbolic reality. Misguided attempts, to be sure, but it is rarely self-evident how to appropriately bring a distortion to flatness.

G: You're theorizing that, while physical reality always remains flat, an abstract symbolic reality becomes distorted and remains so—in spite of maneuvers to induce an artificial flatness—until a natural tendency toward true flatness finally induces real change.

H: Yes. Suppose we contrast our neurotic patient here with a wild animal—a bear. The bear doesn't have a conscious mental model of itself. It has neither ego nor conscience. The bear never represses anything. But what do we mean by repression? We mean that the structure of the conscious self is such that it won't admit into the conscious processes any contradictory urges or thoughts. In this sense, repression isn't an active behavior, it's simply a result of a flawed design of the conscious self. The bear doesn't have a conscious self, which is essentially a mental model of *who* and *what* that is couched in linguistic symbols. So it always manifests its unconscious urges and thoughts immediately, it puts them into action. It never loses touch with the cycles that keep it flat. As soon as it gets hungry it begins searching for sustenance. When it's tired it finds a place to sleep. When it wants to copulate, it goes into a mating mode and copulates. It doesn't defer gratification, or sublimate, or resist its impulses. That doesn't mean that it's stupid, and it doesn't preclude using complex strategies to get what it wants. It does mean it lives in the present tense. Like physical reality itself, the bear immediately makes necessary changes in itself so as to retain flatness.

G: And if our neurotic patient did the same, he'd not be neurotic.

He'd never get so out of sync that he needed to add a ritualistic rhythm to his symbolic reality to flatten it.

H: That's the theory. Freud's argument was that the trauma had to be integrated for the libido to flow freely, and I'm only remarking that his conceptual paradigm is equivalent with that of distortions in a personal symbolic reality. Trying to operate a mental model of oneself that contains distortions is like trying to drive a car with a bent camshaft; you may be able to add counter-rhythms that make the engine's performance smoother, but it will never run efficiently that way.

G: Libido flows freely in a well-designed and well-tuned engine, one in which all the rhythms are integral.

H: Well put. Because we can distinguish between natural and unnatural rhythmic behaviors. Eating, at an elemental level, is a rhythmic behavior, but it's integral to the physical design of the organism. So is sleeping. Ritualistic behavior is not. Ritualistic behaviors are non-integral components fitted into the mental design to make it approximate flatness. That's why ritualistic behavior is so heavily symbolic. Religious ritualism, sexual ritualism, psychopathic ritualism—all are laden with symbols rich with abstract meaning. The physical changes effected by such behaviors have minimal significance. In Holy Communion you're eating and drinking, but the physical gratification is peripheral. Lady Macbeth's hands actually are getting cleaner as she washes them, but not much. Sexual rituals generally climax in orgasm, but the physical relief is only a small part of the satisfaction. Because, the needs addressed aren't to be found in physical reality, but instead in the mental models drawn of it. That's where the distortions lie, and that's where adding heavily symbolic rhythms can induce flatness.

G: But, the mental model is a representation of the physical reality.

H: Never a completely accurate one. If you ask a person to give you in words a full description of who he is and what events shaped him, he will assuredly leave out some important facts. Not deliberately, not consciously. It's just that this mental model he has assembled to represent himself is to a great degree makeshift; it only roughly matches what has really happened and who he really is. Now, if you're sharp you can notice flaws in his recitation, and raise pointed questions that expose the flaws, or distortions. That's the therapeutic process in psychology.

G: A therapeutic process that you, presumably, endorse.

H: Without much reservation. A friend of mine once observed that she no longer had any interest in talking to anyone who hadn't been through some psychotherapy. Because, she said, the conversation was otherwise full of pitfalls, and paths that led in circles. She was right, of course. If any progress is made in therapy at all, the patient gains a recognition that the mental model is not him, or her, that it's only a model, and an imperfect one at that. Which is a tremendously liberating discovery. And you can have a decent conversation with someone who's made that discovery.

G: I guess this problem of distortions carries over into other symbolic realities, the non-personal ones.

H: Like economic models? Yes. The one we mentioned previously is a case in point. Should you have asked an economist to describe how our economic system worked, he'd never have mentioned the accumulating garbage. Foreign debt, employment levels, the Federal Reserve, free markets, government regulation—all that. But he'd have forgotten the garbage. Listening to his recitation, if you were insightful, you might have put your finger on this flaw. You might have said, "Hmm… all this producing and consuming, what happens to the leftovers? How are they recycled into a supply of raw materials for new production?" And the economist would've gotten flustered, and hemmed and hawed and squirmed like a discomfited client in therapy. What makes these distortions so hard to see from the inside is that the abstract model is compensating for them. It has an ingenious technique for concealing the flaw, the distortion.

G: Such as assuming that in the future the problem of garbage will be solved. Likewise with the nuclear waste problem. Technology will take care of it, eventually.

H: Or, with the obsessive-compulsive, using a ritualistic behavior to erase the distortion, from time to time. After which he can reassure himself that everything's okay again. If you don't mind being irreligious, you can argue that the ritual of Holy Communion is designed to flatten a flaw in the meaning-system that is at the heart of Christianity, or any religion, for that matter. The flaw, you could argue, concerns the unanswered question of exactly how the divine essence interacts with physical human beings. You could argue that the rituals of transubstantiation disguise this flaw.

G: Farfetched.

H: But, all such ingenious techniques as we've considered have a common dynamic. In every case of artificially flattening the model, a duration of time is being inserted into it here or there, so that it will seem flat.

G: This is the dynamic you referred to in the analogy of the tree-trunk cross-section? Where you "leaned over and looked down on it?" So that the distortion is no longer visible?

H: And what you're doing then is monkeying with the temporal dimension, which you had supposedly collapsed, across the board, into a single moment. At certain points in the plane you've allowed duration to represent a single instant. It's cheating.

G: Yet an abstract model can get away with this deception. For how long?

H: Not very. The abstract model doesn't exist in a vaccuum. Transcendental or esoteric or hyperconscious as it may appear to be, it is grounded in physical reality, where cartoon characters can't hang suspended in space.

G: I'm fascinated by this idea that mental health is synonomous with flatness in the personal symbolic reality—the subjective point-of-view on the self.

H: The keyword in mental health is *integration*. Integrating means building a symbolic model that is a description of the past that is flat against time. When the self is integrated, there's an ongoing serenity. Energy isn't wasted, libido isn't blocked, and a person flows through life, instead of battling his way through life. He flows along with the cycles that make him up. At no moment is it necessary to include in the model a duration of time. At that moment, whichever moment it may be, the model is already flat. Obviously, very few human beings today are always at peace with themselves. There is always some-thing to accomplish first, something to finish, somewhere to go, and when we make our self-models flat we manage it by including that coming accomplishment. But even when we accomplish it, we're not happy. There's something else. And it just goes on and on, with pre-cious few moments of serenity, until death.

G: Certainly we shouldn't all strive to be Buddhist monks.

H: But our need to *do* should come from natural contingencies, rather than from distortions in the mental models. Flatness of the symbolic reality doesn't preclude active behavior. It just predicates it on condi-tions in the physical reality. You can be utterly serene, and yet hurry outside when an earthquake shakes your rafters. To be integrated im-plies inner peace, not passivity. The Buddhist monk isn't inactive. He moves with the changes in the outside world, responding to them according to what they are, and not according to distorted meanings he attaches to them. It's wise to remember that the most inactive per-son is the catatonic schizophrenic, whose subjective model is so dis-torted that he can't function at all in response to the outside world.

G: It's wise to remember that we don't reside in rural Tibet, either, where it's somewhat easier to achieve daily serenity.

H: Oh, free market forces do their best to demolish serenity. Serene people aren't very good consumers, from the view of corporate adver-tisers. Which serves to point out how distorted their frame of refer-

177

ence is. The problem of distorted symbolic realities runs top to bottom in our society, from the abstract model of capitalism down to the lowliest consumer, and includes just above everything in between.

G: Unirealism applies to everyday life, then, because the distortions in our abstract symbolic realities are responsible for our anxieties, our injustices, and our unhappinesses.

H: Yes. That's the "So what?" answer. Whatever meanings are found in life derive from the relation to flatness. It is only by comparison to this referential standard that values can be assigned to things, to events, to relationships, to actions.

G: What is it about distortions that interferes with efficient functioning? An engine with a bent camshaft doesn't run serenely. Why not?

H: Do you recall how we talked about energy being potential when it was balanced in dimensional complexes? Energy vectors are potential when they're trapped in symbols, and kinetic when they're free? In a perfectly flat frame of reference one doesn't notice the energy. Energy is visible only when one disturbs the frame, and energy is released kinetically. The potentiality is what makes the frame serene, silent, and seemingly eternal. The cycles through which the potential energy flows are exquisitely arranged so that the symbolic reality seems to stand still, despite its composite motions. Look at the night sky. Once in a while you see a shooting star, but that's it for noticeable kinetic energy. It's like a well-tuned automobile engine, with no stress or strain as it runs: the auto mechanic's version of esthetic beauty.

G: No vibrations, very little noise, no visible exhaust, and also maximum fuel efficiency. The equivalent of psychological serenity. And it has far more potential energy than a badly-tuned engine.

H: Yes.

G: So, in a flat frame of reference—a non-distorted symbolic reality—the energy vectors are still there. It's just that they are curved about each other in cyclic balances. So well balanced that the energy vanishes into a silent potentiality. As it does in a spinning top. Like with the electron orbiting the nucleus, you have a poised system of motion where no energy is wasted or lost. And at least with the electron, it can keep it up forever.

H: Exactly. However, if you set up a string of these atoms and trigger a flow of motion that is unbalanced, you will waste energy, through the resistance along the way. That resistance drains the energy of the flow, dissipating it as heat in the wire along which it's being conducted. Heat, of course, is vibration. It's the appearance of unbalanced energy vectors.

G: And such heat-loss is, if I learned anything in high school physics, entropic.

H: Wonderful! And now you see. The system that isn't flat is slowly entropizing. Its potential energy is gradually dissipating, becoming kinetic in the form of heat.

G: Ah, a thermodynamic definition of human misery. But here's the rub—how does one go about achieving flatness in abstract models?

H: The trick isn't in flattening the model. That will take care of itself. You can't distort a model for too long, because it is constantly losing energy and ultimately it will collapse. A new frame of reference, a new model, will emerge in its place, and it will be flat. Rather, the trick is to avoid distortions arising in the new frame. Nature flattens frames. It's our conscious behaviors that unflatten them.

G: This necessary "collapse" sounds rather ominous. Do you mean it literally?

H: Well, as with the garbage piling up in landfills, and toxic chemicals seeping into groundwater, sooner or later physical reality reminds the abstract model that it's inadequate. Collapse is meant literally, yes.

G: But why is it the situation has to worsen so much that collapse occurs?

H: No frame of reference, no symbolic reality, can remain static. Time passes and observations are made, and the frame must absorb them, so that it reflects the passage of time. In physical reality the change is continuous. The frame never loses flatness. That is prohibited. But these abstract models we're discussing try to remain static. That's how the distortions occur in the first place, little by little as the frame resists change. Whether it's the self-model—the ego, or the society model—the state, there's an institutionalization of identity that avoids continuous accomodation. So that change builds up until it has the force of a tsunami, and the distorted frame can no longer cohere. Occasionally it happens with people; they have nervous breakdowns. Usually it doesn't. The ego is protected only by mental structures we call "defenses." Most of us simply suffer mini-nervous breakdowns, and those give us modified flat frames. Other abstract models aren't even that accomodating. The economic model has very substantial institutions of identity. We're talking granite: laws and courts and regulatory agencies and corporations. A model like that can resist change for decades, even centuries. And when the change finally arrives, its cumulative force is devastating.

G: I surmise that you don't approve of America's economic system. You did mention the garbage and the toxics.

H: Our economy perhaps doesn't have the obvious distortions of czarist Russia, or monarchical France, but our model is full of distortions, and all the tricks we're playing to give the illusion of flatness are eventually going to worsen matters. I take the Marxist view here that whatever problems we have are fundamentally economic. Economic energy is the lifeblood of a society. And, it is economic profit that encourages the distortions. Pollution, resource waste, and abuse of the environment are all profitable in the short run. That's the only reason they are permitted to continue. But those three factors are entropic, and sooner or later comes the collapse of an entropizing frame.

The collapse may have a political face, but those three factors usually lie behind it.

G: Are you suggesting a coming revolution in the mode of the French or Russian?

H: Ultimately, yes. I hope it's less bloody. More along the lines of a British revolution.

G: But Marx was wrong, and you could be, too.

H: Marx was too simplistic. I don't know about wrong. Anyway, America has already had three political revolutions in this country. The first was in 1776, but there was a second with the Civil War and and third with the New Deal. Why shouldn't we have another?

----Seven----

Grana: Many historians believe that civilizations come and go in cycles, but while they've hypothesized certain stages in the rise and fall of a civilization or empire, no one has successfully explained a causal sequence to account for the stages. But this idea of frame collapse—of a brand new flat frame of reference growing more and more distorted until it collapses—might readily be applied to the cycle of a civilization. Before we go into that, let's review the dynamics of change in an abstract symbolic reality. Where shall we start?

Heidrich: First question, how does a model or frame become distorted?—the answer being that abstract symbolic realities resist continuous change. Second, why can't a distorted model or frame remain so?—the answer being that it dissipates energy in sustaining the distortions. Third, what occurs when those distortions become too taxing?—the answer being that the old model literally disintegrates and a new model appears to replace it.

G: These dynamics apply to all symbolic realities: psychological, economic, political, epistemological, societal, etc?

H: Yes.

G: And the progressive changes hinge on the state of energy in the symbolic reality?

H: Yes. Which is to say, whether the energy progressively becomes more kinetic, or remains potential. Potential energy is flat against time. Kinetic energy is not. That's why it has noticeable motion. The distinction, of course, depends on point-of-view. The point-of-view that counts is the point-of-view implicit in the frame of reference. Let me demonstrate this with an example. The moon has great potential energy; it's sitting up there resisting gravitational pull by the earth. If it fell to earth you'd realize how much potential energy it has, because it would devastate the globe when it hit. And in fact it is falling; it's falling around the earth. Its motion seems to be kinetic, from one aspect. If an asteroid happened into its path you would witness how kinetic the moon's energy is—the asteroid would be pulverized. But, in the larger view of the moon and the earth together in a system and revolving around each other, there is no kinetic energy. It's all potential.

G: The kinetic-ness or potentiality of energy is relative to the frame of reference you're using.

H: Right. The universe as a whole has no kinetic energy. It's one vast complex system of cycles permeating space and time. But if you take a quick-shot photo of the universe at a specific moment in time from a specific location, you will see what appears to be a lot of kinetic motions—photons radiating, supernovas exploding, galaxies spinning, and so on. The photo is a point-of-view. It's you, it's me. And our point-of-view reveals to us a frame of reference, a model. That which we see as static contains all the time that has passed. What we see as kinetic is time that *is* passing. It is in the process of being integrated into the interference pattern. We can't see the time that hasn't passed—i.e. the future. It hasn't yet materialized.

G: This is terribly difficult to visualize.

H: I know. But here's what I want to make clear: a point-of-view embraces a certain interference pattern, an interference pattern that, as time goes by, evolves. It evolves by integrating into its pattern, through observations and symbolizing, more and more new cycles. The integration process is one of balancing the dimensional values "loose" in the world, which seem to be kinetic. Okay? And once these dimensional values are balanced, they become potential. As integral balances in the interference pattern, they stand still. Whatever motion is *in* them doesn't cost the system anything to maintain. The system is not being diminished entropically by the energetic expense of keeping all the parts moving.

G: Although the second law of thermodynamics says it must be entropizing.

H: To an observer, there appears to be a steady entropic loss. After all, he's making observations, and observations are the conversion of potential to kinetic energy.

G: Ah. Since he is assuming that the frame of reference is static, it seems to be crumbling as he looks at it.

H: Yes. Entropy is an illusion peculiar to the viewpoint that is slowly becoming obselete. The frame seems to be running down. But the new frame of reference will harness all the entropic motions.

G: Therefore...if I could be an observer outside the universe— hypothetically—no matter what point of time I looked at, there'd be an interference pattern there. Which would contain all the passed time. And, if I compared one of these patterns with a prior pattern, I'd see entropy from the earlier one becoming incorporated into the later one. That would be time going by. And that would be the process of transformation between them. In short, entropy is part of the process of change. It is a loss only relative to the prior frame.

H: Yes, relative to the prior frame. And that is where the observer is! He is steadily moving into the future, but his model of what's going on lags behind. He *does* see an increase in entropy. He sees his frame of reference being slowly eaten away by heat loss. He sees a gradual deterioration of form into random motions. From this he infers that all systems eventually run down. He formulates the second law of thermodynamics. But, he's making a judgment based on the conceit of his own self-importance.

G: Physicists will feel the sting from that.

H: To paraphrase Fred Alan Wolf, who is a physicist, observation produces information and entropy. That is the operative transformation here.

G: But, producing information and entropy isn't quite the same as building a new frame of reference, is it?

H: Information is symbolic, true? And what eventually happens to information? It gets integrated into a new point-of-view, right? What's changed between the point-of-view you have now, and the point-of-view you had at age twelve?

G: I've made observations about life, formulated them as symbolic information, and integrated it into my overall perspective.

H: So when you exist as a point-of-view—which is to say that you make observations—you're absorbing time as it passes. And the blueprint of reality is that simple.

G: The upshot is that both information and entropy are temporary. The new point-of-view has integrated them back together again, in its new interference pattern.

H: Yes. The symbolic information melds with the entropy in a way that makes the symbols integral to the new frame of reference. So that neither the entropy nor the information is visible anymore. The new frame just takes the synthesis for granted. In the same way that you, during the day, take for granted all the symbolic information you've assimilated over the course of your life. It's part of you now. It's integrated into your self and your identity.

G: Nevertheless, even though I assimilate information and entropy through my life, so that my point-of-view is always evolving into a newer one, I will eventually die. Despite my best efforts, the frame of reference that is me will eventually collapse.

H: And by the time it does, you will have procreated, one way or another. You'll have taken the symbolic components of your point-of-view and randomly mixed them with the symbolic components of others, to make new frames of reference. Children. That method of self-replication is far more effective than any that would keep you corporeally intact. It allows tremendous flexibility of response to environmental change, and corrects chromosomal damage.

G: But is it inevitable that a symbolic reality—such as a human being—must eventually collapse?

H: Oh, I think so. We must remember that in reality, anything that comes to exist immediately begins curving into a cycle. As with the roadrunner, shooting off into space. As with the photon, curving gradually through gravitational fields. Any kind of motion through space and/or time becomes cyclic. In order for this to be a zero-sum universe, any phenomenon manifested within it must curve cyclically. In a self-enclosed universe there can be nothing that is absolutely straight.

G: But isn't there some distinction between the cyclic motions in an atom and the cyclic motions of reproduction through the generations?

H: Only from point-of-view. In any case whatsoever, the cycle is of form moving in space, through time. The moon circling the earth arrives at its starting point a month later than when it left. Likewise, between your birth and death there's some duration of time. If you could be entirely objective, you'd see that all cycles are the same. But you can't, because you're a point-of-view, and a point-of-view is constructed entirely out of cycles. You tend to believe that you're solid and fixed, and moving along with time as it passes. But that's part of the illusion. The solidity and fixity you perceive is there because your point-of-view is an interference pattern of cycles.

G: Then just as the dimensional values themselves curve with time, any symbolic form made out of them also curves with time.

H: Sure, sure. Stars, galaxies, people, civilizations. The curving is marked by entropic disintegration, as the universe slowly evolves as an interference pattern. And not only concrete things, but abstract representations as well: theories, relationships, cultures, languages, and so on. Look how habits develop. An activity becomes cyclic, so that it repeats over and over. But even the habit itself curves into a cycle, so that one day you suddenly notice it's a bad habit and break it.

G: And what about the eternal verities? Truth, love, integrity?

H: According to the principles of unirealism, such terms describe the symbolic transformations themselves, and as long as our universe has a single blueprint, the power of such concepts is immutable and invincible.

G: Okay, we've digressed again, and I'd like to return to our jumping-off point. We've established that symbolic form inevitably curves in cycles through time, and that insofar as such form composes a frame of reference, that frame will also go through a cycle at the end of which it collapses. Furthermore, that from the ashes will rise a phoenix—a new frame of reference that harnesses the entropic disarray that sabotaged the prior one.

H: And that such progressions characterize not only human beings and what they create, but also the universe as a whole.

G: A point with which cosmologists would disagree.

H: I'm afraid so. I find it curious that physicists who believe the universe will wind up as total entropy have no difficulty embracing the idea that a new civilization can spring from the wreckage of a ruined one, or that a species might mutate in adaptation to environmental change. New form emerges all the time, everywhere, spontaneously. But physicists like to think that their physical reality operates under a different set of transformational laws than those that apply to life-forms. This would be forgiveable, except that they can't even explain how—by means of their physical laws—form came to exist in the first place. How did the chaos immediately after the big bang spontaneously develop into a cosmic foam of spiral galaxies and solar systems and human life? It shouldn't have happened, you know. So I take issue with their assumption that there are two different sets of transformational laws. From both logical and intuitional standpoints, I find that assumption unnecessary and unesthetic. A single set of transformations is more beautiful and more elegant than two sets. Hence is it likely more truthful.

G: Anyhow, it's academic, the ultimate fate of the universe. We'll all be dead before we can prove that entropy doesn't win out.

H: I suppose so. But so far as our abstract models are concerned, the question isn't academic at all. Entropy is an illusion subjective to the frame of reference. And the snapping back into flatness does harness entropic motions.

G: The frame of reference that perceives entropy increasing is distorted, is that correct?

H: Yes, in the sense that it's living in the past. Such a frame doesn't necessarily have to be conscious of the entropy, as in, "Hey, look at all that entropy—I'm disintegrating!" But it will certainly experience the draining effect of the entropization.

G: Okay. Now, as I understand this, when a frame is entropizing the potential energy in it is being gradually made kinetic. The balanced cyclic components of that frame are losing their equilibrium. And, I guess you could say, falling apart.

H: Those balances, of course, are of motions. Because they're balanced they seem to be in stasis. Like a rapidly-spinning top, or gyroscope.

G: And the top, as it runs down, suddenly starts jerkily flying this way and that. The energy in it is revealed, as balance is lost. Or an atom smashed in a cyclotron, with subatomic particles flying off in various directions.

H: And so with the frame of reference. As it entropizes, you see revealed all its hidden energies, radiating away as vibrations and waves.

Vibrations and waves, of course, are unbalanced dimensional values. They are kinetic energy—kinetic to a particular point-of-view.

G: Like the solar system. As a whole, its composite motions are potential. But if I flew a spaceship into the path of Jupiter I'd be making, in some crude fashion, an observation. Converting potential into kinetic energy as the planet plowed into me.

H: Yes. But suppose you were some gargantuan being, whose cellular structure was composed of stars and galaxies. Then the solar system would be purely potential energies, nicely balanced in the same way that atoms and molecules are in your own body.

G: I've got it now.

H: Our topic was the ontogeny of civilizations, and how they crumble at the end of a cycle. The rise and fall of the Roman Empire, for instance. An empire—or any kind of society, for that matter—is made up of cycles. Birth and death cycles, production and consumption cycles, seasonal cycles, cycles of successions and elections, cycles of religious celebrations, and such. An infinite number of rhythms go into the illusion of a static society. It appears to change very little from day to day or year to year, even though everything in it is changing all the time. It appears to remain the same because the changes are occurring in rhythms that are balanced. Right down to the most discrete rhythms: the biorhythms, the diurnal and monthly rhythms, of every individual. Everywhere in the society, the observer can perceive rhythms, but there is a collective effort to keep them all in harmony. There's some good research being done these days on the rhythms hidden in simple conversation, with each participant unconsciously entering into a subtle dance of speech and gesture. Likewise, there's evidence that two women sharing a home find their menstrual cycles becoming coincidental. At all levels, from the biologic to the conscious, citizens of a society sense and harmonize with rhythms that, while balanced, permit them to have a coherent social structure.

G: The society is a frame of reference, then. A symbolic reality, an interference pattern of balanced vibrations, rhythms, and cycles. Just like the physical universe itself.

H: However, unlike the universe, which stays flat, the society becomes distorted. And then the potential character of the cycles deteriorates into kinetic energies.

G: The cycles break down?

H: Ultimately, they break down, yes. Initially they only distort. If you imagine a cycle as a perfectly round circle, then in distortion it becomes elliptical. If the cycles actually break, well, that's the end of the society. You can't have a society where no new children are born, or where production stops, or where there are no new leaders chosen. But the cycles can be distorted, so that the overall interference pattern isn't flat against time. And when that happens, there is entropic loss. If a cycle is distorted, energy must be expended to keep it in balance. That energy expended is an entropic loss.

G: Okay, let's talk about a society with which we're familiar, as an illustration of the dynamic.

H: A prime example is the Soviet Union. All those Soviet people standing in lines for much of their day. There's an entropic loss. There's a society stretched to the breaking point by distortions of what ought to be natural cycles. The free market should flatten the frame with its so-called invisible hand. But it can't, because dogmatic beliefs, institutionalized in a bureaucracy, preclude such adaptations. In a healthy society the cycle of production and consumption is well balanced. Little energy is lost or wasted. All the energy remains potential.

G: Although all the members sooner or later die.

H: Which demonstrates the point. From one citizen's point-of-view entropy is occurring. He feels himself getting older, slower, wearier. Then he dies and goes to dust. But a society can cohere for thousands of years with virtually no change—barring severe climatic shifts. So, from the point-of-view of the society, there's no running down. The old man is replaced by an infant, and on and on it goes. The energy of the cycles remains potential. Nothing is lost. There is no entropy.

G: And in the Soviet Union the production-consumption cycle is distorted?

H: Applied Marxism has corrupted the integrity of the cycle. It severs the relation between production and consumption, and inserts this horrendous bureaucracy between them. The bureaucracy is observational. It plays at creating information. But creating information means simultaneously creating entropy. And instead of integrated energies forming balanced cycles, there's energy running amuck. The government tries to control it, to force it back into properly balanced rhythms, by more central planning. And that's theoretically impossible. You can't achieve a balance that way. So, the people spend even more time waiting in line to buy bread. Standing still in a queue is, by our definition, random motion. Entropic. So eventually the socioeconomic distortions must make the political controls give way.

G: Why? They seem to be managing.

H: Oh, and so did the czars. For centuries. As did the aristocracy of France. But the entropic losses are cumulative. Poverty deepens. The necessary police force grows in size. The gap between rich and poor, between privileged and unprivileged, widens. You're sustaining distortions in the frame, and the energy to sustain them has to come from somewhere. It comes, ultimately, from the healthy cycles in the frame. The rhythms and cycles that make it worth living: family life, creative play, intimacy. These get plundered for all the potential energy in them. You have a socioeconomic system cannibalizing itself.

G: But what causes this sort of thing to get started?

H: Consider a simple agrarian society. Agriculture is visibly cyclic. When you grow crops on a plot of land, you must restore to it a fertilizer equivalent to the foodstuffs you harvest. If you keep a simple vegetable garden, you mulch. If it's a larger field, you plow under the stubble and you rotate crops and you give it fallow years. You fertilize it regularly with the dung from your beasts of burden. With such basic husbandry, you can use this land almost indefinitely. There is little energy loss, little entropy. And your production-consumption cycle is balanced. In primitive cultures human waste is used for fertilizing the land from which comes food that feeds people and nourishes them before becoming human waste. The only element missing in this cycle is that of the dead bodies of the old people. Technically, these corpses should be chopped up and used as fertilizer, as well, but it's only a marginal loss when they're buried elsewhere.

G: So what you have is an undistorted model, this agrarian society.

H: An economist or physicist will look at what's going on and argue that there is plenty of kinetic energy being expended. You can see the oxen trudging up and down the rows, you can see the husband and wife making love, and so on. But as a whole, it's a very flat frame. It is sustainable for centuries—millenia, maybe—with the societal members in reasonable comfort. Peace of mind, satisfaction, sense of belonging—these are experienced only in a flat frame.

G: How would you go about distorting such a frame?

H: If you want an agricultural distortion, you might start by analyzing the cycles. You figure out that what's being returned to the land by fertilizing is nitrogen. And it so happens that there's a nearby source of nitrogen. So you figure, why go to all the trouble of fertilizing organically when I can just dump a load of nitrogen on my acres every spring? It may not last forever that way, but by then I'll be dead anyway. Or, you may invent a tractor, which can be a lot less hassle than a team of oxen. But you can't fertilize your land with tractor exhaust, can you? Well, no matter—with what you profit by making yourself more efficient you can buy a new plot of land when that one's barren.

G: And that's the kind of thinking that goes on in distorting a society?

H: It's corrupt thinking. It's corrupt, and it's corrupting.

G: It sounds rather like ordinary conscious thinking, though.

H: Well, yes, it is. You're extracting symbols from the flow of reality and using them informationally, inserting them back into reality to manipulate how things change. That's the conscious process, in a nutshell.

G: And you're opposed to it?

H: Not *per se*. But consciousness is a technique, and we ought to understand what we're doing when we employ it. Otherwise it can be terribly destructive. Symbolic manipulation needs to be executed with extreme sensitivity to the world. Consciousness demands an ethos.

G: In the case of the farmer, he is manipulating symbols when he recognizes the value of nitrogen and then uses it for his own purposes?

H: Yes, you can say that. But he was also manipulating symbols when he used animal dung. The difference is that now he's abandoning the ethos that linked him to the land, and he's become a businessman. This isn't meant as condemnation of the farmers and ranchers of America, who have virtually no control over the system that demands cheap food and gives them little in return. But good grief, the continents are losing billions of tons of topsoil each year through erosion—erosion directly the result of corrupt resource management. That is entropy, pure and simple. It's the same as the Soviets queueing up in long lines to get ration coupons so they can line up again to buy bread. In both instances the entropic loss is a consequence of symbolic manipulation shorn of any ethos. In both instances it's a matter of consciousness in ascendance over awareness. And, in both instances the loss is ominous. The fact that there's such entropic loss indicates that the model is grotesquely distorted, and forbodes a coming cataclysm.

G: Perhaps people just don't see it like you do.

H: Oh, everybody knows these models are distorted. No one has to tell you that standing in line five hours for a loaf of bread is a waste of time and energy. People just don't know how to correct the distortions, or they're afraid of the personal consequences if they do try. And with some justification. So everyone makes compromises with the system, and ends up being a little or a lot corrupted. Which is why societies get deeper and deeper into the quagmires from which they can never escape.

G: Without collapse.

H: Yes. This isn't just the Soviet Union we're discussing here. It's America, as well. It's every society that's expanded and collapsed in the history of the planet. Egypt, Persia, Greece, Rome, Islam, the Ottoman Empire, Spain, Germany… Each began as a relatively flat model, a reasonably healthy and integrated society with enormous potential energies. Thereafter they gradually progressed into such incredible distortions that they were vulnerable to revolution, fragmentation, or conquest. And one of the clearest evidences of this is the fact that in all of them, the final stages were characterized by widespread and wholesale corruption.

----Eight----

Grana: Can we distinguish between acting corruptly and acting with integrity?

Heidrich: The word *corrupt* comes from Latin, and the original meaning was "to break into many parts." *Integrity*, on the other hand, is derived from integer, which means "one"—a unity. So, *corruption* and *integrity* have contrasted meanings, which refer, respectively, to treating the world as a bunch of disconnected parts, or as a holism. With corruption, the tacit assumption is that the many parts are connected only by causal chains of events. Which, according to unirealism, isn't true. Acting with integrity, on the other hand, assumes that the world is holistically unified, and that the individual is intimately and implicitly connected with the whole. Given this type of connectedness, the individual can allow his behavior to be governed by the here-and-now demands of the world. He will know what to do because he is inherently knowledgeable, by means of the connections, about the world. And, by responding to the here-and-now, he maintains a flat frame of reference. Acting corruptly involves a denial of such mysterious connection, and a reliance entirely on apparent causal connections. No one can keep a flat frame of reference that way. And so the corrupt individual must find artificial ways to keep his frame flat. He depends on transtemporal symbols, using them to give the illusion of flatness.

G: For example, transtemporal symbols of exchange? Money?

H: Oh, yes, money. Money. If people could see what they're doing when they make a big deal about money, then money would become nothing more than a joke. Probably the biggest joke in history. In fact, the obsession with money is such a singular symptom of psychic dysfunction that I should think people would be embarrassed to let it show.

G: Let me in on the joke?

H: Okay—you want a flat frame of reference, right? Even if you don't know that's what you want. The desire is instinctive. A flat frame is more serene, it is more efficient, and human beings are designed to seek and find efficiency.

G: How can you apprehend when your frame isn't flat?

H: You can feel it emotionally. Chiefly by anxiety.

G: Anxiety is the sensation of distortion in one's frame of reference?

H: Sure. For both the individual and the society.

G: Hmm.

H: Now, when a frame isn't flat against time, it's spread out over time. Here's a visual image of what I mean: a flat frame is like a piece of silk, while a distorted frame is like a piece of terrycloth. In the terrycloth, the nap sticking up represents the distortions. Height represents the temporal dimension. The silk, in its smoothness, has no height. The terrycloth, with its nap, does.

G: Since height is time, the terrycloth is spread out over time, compared to the silk. All right, I can imagine that.

H: When we say that the frame isn't flat, we mean that as an interference pattern it isn't properly balanced. Its constituent motions, which are flows of energy, have become kinetic. Subjective perception of this imbalance is felt as anxiety.

G: We're talking about an abstract mental model here, but I presume there is a metaphor in the homeostasis of the body. When the rhythms that comprise homeostatic balance are imbalanced, you feel sick.

H: Yes. And yes, this is a mental model we're referring to, the sum of conscious and unconscious thoughts, attitudes, personality, and so on. Homeostasis in the more physical self is indeed a fair metaphor.

G: By the way, if you are to look down on the terrycloth surface, you won't see the distortions, am I correct?

H: Yes, because you're eliminating the dimension of time that way. You are looking *down* time, so to speak. But that's an objective act. You may fool your conscious self with it, but you're not going to fool the rest of your mind, which still feels unease at the distortions. And, if the condition is extreme enough, the anxiety will be expressed through some biological malfunction, or through self-destructive acts.

G: That makes sense.

H: But, money can be used to artificially flatten the frame. It's only a parlor trick, but an amazing number of people convince themselves it's real magic. Here's how it works. What is money? It's a medium of exchange, symbols that are intrinsically worthless, but assigned an extrinsic value according to what they represent. And what do they represent? Commodities with intrinsic value, commodities which can satisfy real needs. Money gives the one who possesses it the power to satisfy needs that aren't even present in the immediate moment. Money here now will buy bread tomorrow. Money here now will guarantee that you won't be insulted or rejected next week.

G: I'd have to agree with that. Money does assuage anxiety, in just that way. So you're saying it makes the distortions seem like they're not there, since anxiety is how you sense the distortions to begin with.

H: The objective act of looking down at the terrycloth so that the distortions vanish? Money performs that act for you. And, it can do so because it has a transtemporal mobility. It can readily represent satisfaction of needs in a time-slot other than the immediate one. By mentally juggling such symbols, you can make nearly any frame appear flat, past or future.

G: But aren't the distortions in the present frame?

H: Yes, and that's why it's only a parlor trick, and not real magic.

G: But why flatten frames that aren't immediate?

H: When a non-conscious creature feels anxiety, it responds with a behavior that gets rid of the source of the anxiety. It doesn't know how to maintain a distorted frame. It fights, or flees. If it can't do either, the anxiety can literally kill it. It will die of fright. Conscious animals like human beings don't usually directly attack a source of anxiety. Nor do they flee. And they very rarely die of fright. Instead, they project themselves—via transtemporal self-symbols—into non-immediate time-slots, and operate on the present from those places of safety. That doesn't get *rid* of the anxiety. If you're afraid of starving in the present, you'll more than likely be afraid of starving in whatever future time-slot you imagine yourself into. But with money, you can satisfy the hunger in that future time-slot. It is, however, a sign of dysfunctioning to live in non-immediate time-frames. Visit them, yes. Live in them, no.

G: And you believe that people with strong acquisitive urges reside in such non-immediate time-frames.

H: Back up a second and let me qualify something. Rather than categorically calling it a dysfunction, I would note that it can be an adaptation to perverse contingencies. If you're in a penitentiary and you dwell in the past, or the future, that might be safer for you, and healthier. If you're writing a novel, and dwell in a fictional place with fictional characters most of your day, that is probably necessary. But the inmate and the writer know perfectly well what they're doing and why. They are making choices when they do it. People with strong acquisitive urges—let's call them greedy people—don't have any clue what they're doing, or that they could do things any differently. They *never* live in the immediate frame. They don't even know how. All they see of the present is opportunities. Their selves are stunted, and their egos are inflated. They actually think that they're smart because they can steal from the less acquisitive. They're like little children inside. Albeit rich little children.

G: Perhaps we should clarify the distinction between *self* and *ego*.

H: We can postulate that the self is the real you, while the ego is the self represented with cognitive symbols. That is, the symbols used for thinking and communicating. So the ego is an abstraction of the self. How accurate a description it is, how amenable to correction, is a major criterion of one's mental health. Ideally, the self and ego should mesh seamlessly, so that a situation doesn't call for contradictory behaviors. Carl Jung interprets ego as the focal point of consciousness, and the self as the totality of the psychic functions, which are definitions I'd go along with.

G: The ego, then, is the conscious self?

H: Yes, ego is the self skipping about transtemporally through the medium of consciousness. If you wish to imagine a possible situation, you can't imagine yourself in it without an ego to represent your self. Your self is anchored in the present.

G: So where are the distortions?

H: The distortions are in the consciousness, which is an abstract model of the self and the world. But there's no direct method for the consciousness to apprehend flatness. It can know whether it's a good model only by the effect it has on the body, which is where it's grounded to physical reality. The perception of flatness or distortion is an emotional response. Positive emotions generally come with flatness, while negative emotions generally indicate distortions.

G: And using transtemporal symbols—like money—to mute a negative emotion like anxiety allows an artificial flattening of the consciousness model. That technique is what you mean by corruption.

H: Yes. The ease of employing transtemporal symbols makes corrupt thinking almost unavoidable. Only a saint can escape it. Even Christ was tempted.

G: Which suggests that the temptation to corruption is part of our nature.

H: That temptation is the price we pay for our consciousness. To avoid it one must remember that it isn't the future, or the past, which is important. The present is what's important. The immediate. The future is made in the present, in ways that we, with our primitive notions of causality, cannot understand. But if you take care of the immediate, if you fully respond to it, the future will take care of itself. And of you. Castaneda's Yaqui mentor, Don Juan, suggested a sobering reflection. He said you must live life with Death perched on your shoulder. Not as a tormentor, but as an advisor, who reminds you constantly that there mayn't be a future for you. So take care of the here and now. But you needn't look so far afield for advice like that. True Christians are surprisingly good at the same focus on the immediate. They simply trust that God will take care of them in the future, as long as they take care to live with integrity in the present.

G: But if you don't live with integrity in the present, you're vulnerable to corruption.

H: And not just by money. There are many little ways to assure ourselves that our frame of reference is flat when it's not. And ways to harden the model against cracks of doubt.

G: And you consider all such reassurances corrupt?

H: If you have to keep reassuring yourself that you're right, then you're likely wrong. If you're at a job you loathe, and you keep telling yourself to stick it out X number of years and then you'll retire and be happy—well, that's corrupt thinking. It's like saying I'll just steal a lot this one time, and then be able to afford to live an honest life.

G: No, that doesn't work, does it?

H: Imagine some priest in the Inquisition setting a young girl afire for confessing under torture to consorting with Satan. Reassuring himself that he must strike the match because it's God's will. That's corrupt thinking, to me. And then there's the trap of imagining the dreadful consequences. If I don't do this the children will starve. The wife will leave me. The Inquisition will torture and burn *me*.

G: I do hope you aren't proposing that people disregard possible consequences, and live only according to their notion of integrity. That's a dangerous philosophy.

H: Sure, if for twenty years you've been thinking and acting corruptly, and then one day you wake up to the delusion and try to make a stand for integrity.

G: But not if you develop it as a way of life?

H: Look, the world *likes* flat frames of reference. If you respond to the moment, keeping your frame flat, the future you enter into will take care of you. When you align yourself with the immediate, you're aligning yourself with underlying cycles of reality that predate civilization itself. When you tune in to those cycles, reality will guide you, advise you when to eat, when to sleep, when to copulate, when to smile or cry or laugh, and when to speak out. If something needs to be done, reality tells you to do it. All you have to do is learn to listen. And if you listen, and respond, your frame will stay flat.

G: It sounds like you're arguing a logical basis to faith.

H: Logic isn't necessary. Such faith can even be learned heuristically. A good deal of the maturation process is simply learning to do what needs to be done, when it needs to be done.

G: And that, everyday reality doesn't always permit.

H: But our everyday reality is a disaster. We're all walking through a war zone. Which is the long-term consequence of too many people listening too much to ego and not enough to self. Understand, it wasn't always this way. Distant times in human history people did live in the immediate, and they lived well.

G: Not in written history.

H: Written history came with transtemporal consciousness. Written history is the record of corruption.

G: But your faith verges on the mystical...

H: In those forgotten times, the world was imbued with mysticism, and mystical awareness. How did the hunter know exactly where to drive his spear into the sabertooth tiger? How did the healer know what herbs to pick for a compress? How did old women know when to plant seeds? Even today, how do animals in the wild know what they do? Instinct alone doesn't adequately answer for it, because animals can learn new instincts, even when heuristic experience isn't involved. Wild animals—like primitive human beings—know by listening with uncanny acuity to the world. It tells them. And all of us were born with the same ability. We lost it. It gets drowned out by the conscious thought processes.

G: Just how does that happen?

H: Consciousness is the ongoing flow of transtemporal symbols in the mind. But the activated symbols don't come from the immediate environs. Perhaps they're triggered by some element in the immediate, but then the flow carries on and on, raising this idea, remembering that, which then segues into something else. And behind it all lurks the spectre of the ego, defending itself adroitly, blocking off this thought, locking onto that thought, vigilantly preserving a mental model of who you are. If your mind is busy trying to prove that the ego is an accurate model, how are you ever going to hear what you need to? So you don't respond to the immediate. Not effectively, anyway. You're ever translating those preconscious perceptions into conscious terms, and too often the translation is stopped dead because it doesn't fit the routine pattern that's already established. Such thoughts aren't permitted to reach articulation, and even when they do reach it they're rationalized away.

G: A rather harsh indictment of ego.

H: Yes. Well. All the magnificent accomplishments of men and women are achieved during a transcendence of ego. Did DaVinci think about how great a painter he was while working on the Mona Lisa? I doubt it. He simply became a great painter during the act. Does Lendl play tennis with his ego? I doubt it. And the memorable acts of valor, of heroism in the face of death—tell me, how did the ego feel about such acts? The answer is that it didn't. The ego was temporarily overwhelmed. Or tell me this—where is your ego when you're in moments of real emotional intimacy? It isn't active. It can't be. Ego and intimacy are mutually exclusive.

G: But surely there's a place for ego.

H: Of course. Consciousness is a wonderful tool. When the self is directing those adventurous rides into past and future time-slots, into conjecture, into imagination and fantasy and creation, then consciousness is a sheer delight. But when it's in service to the ego, watch out.

G: The temptation to corruption, correct? So let's talk a bit about morality. We haven't even mentioned the concept of moral laws.

H: Anyone who has studied situational ethics realizes that moral laws simply don't work. Generalities are fine—thou shalt not kill, and the like—but they don't apply very well to extreme cases. If there's a moral rule that seems to always work, it's that you do what the situation demands. You do what your heart, your gut instinct, tells you you ought to do. That does not by any means authorize whim and caprice. It means listening to the whole of yourself, and not just your consciousness, because your consciousness finds rationalization of distortions a little too facile.

G: What about principles?

H: They're fine in hindsight. You may look back on your life and draw certain principles from it. You say that you've always spoken out for what you believe in, or that you've never broken your word, or that you've always tried to be honest. But it's idiotic to turn that around and start living according to your description of how you have lived. Why bother living at all, if you're just going to follow a format? That takes all the challenge out of life, doesn't it?

G: And yet you demand integrity.

H: Integrity is a way of doing things. You try at every moment of decision to behave with integrity, but as soon as you start priding yourself on how much integrity you have, you're lost. So you just go on doing what the moment itself calls for. Later, someone comments that you showed a lot of integrity. But you don't deserve great hurrahs, because you just did what reality told you would keep your frame of reference flat. Big deal. Most heroes are uncomfortable with medals being pinned on their chests. They don't quite trust the ego. With good reason.

G: And what about God? There's precious little room for Him in that philosophy.

H: No? Perhaps not God the white-haired patriarch who sits aloof in heaven and watches all this with alternate amusement and anger. But we've terribly insulted God by rendering such a graven image of Him. We've banished Him from our everyday world. We've deprived Him of participation.

G: Are you implying that He does participate?

H: Certainly, certainly. Unirealism says that God is the *way* things work.

G: Then God is not the mechanic, but the mechanics of reality.

H: Succinctly stated. And why not? When we remark that things turn out for the best in the long run, or that every dark cloud has a silver lining, or that with age comes wisdom, what are we doing but complimenting a God who is omnipresent with His ways? Westerners have artificially divided the world into the sacred and the secular, as if in one sphere you live by humankind's laws, and in the other by God's. That's absurd. Reality is the same everywhere. And should we live by moral rules, and say that they're God's commands? That's equally absurd. What, has God lost His voice? Or has He no imagination? No, God is omnipresent and omniscient and omnipotent because he is the *design* of the world. He is the blueprint, and that blueprint is implicit in all the workings of reality. The inexorability of the transformations is His faculty of judgment.

G: An interesting approach to the concept of the creator.

H: The creation is ongoing. The creation is in the moment.

G: I suppose, then, that all the gods men have worshipped are hypostatized versions of this divine mechanics of reality.

H: Once upon a time in our dim ancestry, God was everywhere. His spirit infused the animals, the winds and rains, the seas and rivers, the forests, the stars, the activities of human beings. But with the development of symbolic communication and embryonic consciousness, human beings began separating the spirit out, and replacing it here and there with causality. Over the millenia, you trace the progressive expulsion of God from the world, as the world becomes more and more secular and less and less sacred. The divine spirit became fragmented. With time, each species of animal had a particular spirit, and each force of nature the same. Each activity of human life was governed by a special deity: war, love, work, agriculture, and so on. Eventually these gods were driven up to the top of Mt. Olympus. And, finally, the divine spirit was driven even from the mountaintop. God went into the sky. He became a monotheistic tyrant. Then He even lost His ability to communicate with human beings. He was eventually declared dead because nobody had heard from Him for so long. Unirealism puts Him back into the everywhere. If He's everywhere, if He is in the way things work, then nobody has any right to tell others what God's will is. Everybody must interpret that will for himself or herself. That's the way it should be, if reality is entirely subjective.

---Nine----

Grana: God ceased to be significant because science had found that the world was composed of causal sequences. Even in the time of Newton, who envisioned reality as an enormous clockworks, there was still room for God as the designer of the machinery, who started it off into its perpetual motion. However, since science has traced the universe back to a single big bang twenty billion years ago, there's no more need for God at all. Science effectively killed God—declared Him dead because he had no active part in the universe.

Heidrich: But to a unirealist, God doesn't have an active part, He is the active part. God is not the creator, He's the creation itself.

G: Which would seem to suggest that the concept of God can be reconciled with the concept of causality.

H: I think so. What do we mean by causality? We mean the linking of two events along a time-line, such that the former will always be followed by the latter. No one, no philosopher or scientist or theologian, has ever satisfactorily explained why such sequences occur or what they mean. Whenever we isolate such a causal connection, we've not explained, but described. So, to state that our universe is entirely causal in character really tells us little. We have merely proposed that there are an infinite number of time-line links. We haven't said why. Nevertheless, in a causal interpretation of the universe, it appears that God is irrelevant, because these causal connections are autonomous and absolute. They don't require any motive force or controlling agent. That is why causality drove God out of our lives. Drove Him into the open sky and then banished Him forever.

G: I suspect that you're about to explain what causality really means.

H: Well, according to unirealism, causality is a partial aspect on motion in the four dimensions. A single causal link is like a vibration of which you're only seeing the motion in one direction.

G: When you see a vibration you're seeing a partial aspect? Of what?

H: Of four-dimensional cycles. If you take a cycle such as the electron whirling around the nucleus, and flatten it by removing one dimension, you'll see an oscillation—a vibration in one dimension, back and forth, back and forth. You're collapsing one dimension of the motion, but the motion is still there, you're just seeing it aspectually.

G: I'd never looked at it that way. An oscillation, or vibration, is a cyclic motion minus one dimension.

H: Oh, yes. All the motions that compose reality are nothing but motions in a certain number of dimensions, with the dimensional values relatively fixed. By playing with the number of dimensions, or the values of the dimensions, you get all sorts of aspects. Say, for instance, you have a circular racetrack. The horse's motion on it is circular—cyclic. Reduce the dimensional value of the width a bit, and the track become elliptical. Reduce that same width value down to zero—in effect collapsing the dimension of width—and you'll perceive the horse running all the way to the end of a straight course, at which point it turns and runs straight back. From a cycle, you've made an oscillation. Now, take that straight track on which the horse is oscillating and move it steadily sideways, relative to the background, and you'll see the horse moving in a sine wave.

G: The whole universe is that simple?

H: The interference pattern of the universe is a complex tapestry, but sure, the essential dynamics are simple. Do you recall how esoteric trigonometry seemed, until it was explained that all the curves it described mathematically were nothing but conic sections? Take a three-dimensional cone and slice across it at various angles, and you get various two-dimensional curves: circles, ellipses, parabolas, hyperbolas. That cone is a representation of the universe, except with one fewer dimensions. Just pretend that the apex of the cone is the beginning of time—the big bang, if you will—and that the cone, expanding outward in three dimensions, is the universe, expanding outward in four dimensions. Now, taking that extrapolated four-dimensional cone, slice across it. That slice is a single moment of time. And when you perceive it, you'll perceive a cosmos. A visible universe, in three dimensions.

G: And that is the basic 4/3 dimensional reduction that gives us symbolic reality. Akin to the 3/2 dimensional reduction that gives us conic sections. The difference being that the 3/2 slice really can disregard the eliminated third dimension, while the 4/3 slice must express its eliminated fourth dimension in its symbolic forms, as passed time.

H: Yes. Furthermore, if you take a conic slice perpendicular to the height of the cone, which is analogous to being perpendicular to the flow of time, you're going to get a two-dimensional circle that completely disregards that third dimension of height. In it, the two-dimensional matrix will have right angles. However, if you want to indicate some of the height dimension in the slice, you can do so by angling the slice a bit. Then you'll have a strong hint of the hidden presence of a third dimension. But when you do, you'll have an ellipse instead of a circle. Right? In that ellipse, the two-dimensional matrix will *not* be perpendicular to time. You can pretend that it is, but you'll still notice odd things going on at extreme distances in the ellipse. Analogously, when we perform our 4/3 reduction to manifest a static and visible universe, that universe must include the reduced dimension of time, and when it does we also see odd things going on at the extremes of the Euclidean matrix.

G: You're bringing relativity into this again, aren't you.

H: Let's try another analogy. The Mercator projection is a two-dimensional map of the curved surface of the globe, which is a three-dimensional surface. But there's an exaggerated stretching at the edges of it, so that Greenland, for example, looks huge. This stretching is a consequence of expressing something with three spatial dimensions in just two dimensions. Likewise, when you express four-dimensional reality on a three-dimensional matrix, you also notice stretching. But that stretching is not only spatial, it's spatiotemporal.

G: The space-time continuum.

H: And all these phenomena are tied together: the passage of time, the expanding of the universe, the evolution of form, the constancy of the speed of light, the absence of anti-matter, the relativity of time and distance. All are illusions integral to the symbolic process of creating the universe. And if you fit them all together in an abstruse mathematics that's probably beyond us, you arrive at a zero-sum equation.

G: Which, at a given time, manifests as a pattern of balanced cycles—an interference pattern.

H: And the cycles can be converted into sine waves and oscillations and pulses, if you reduce the number of dimensions that express them. Causal links are discovered when you observe the underlying cyclic structure from a certain aspect.

G: That is, when you perceive cyclic motion as a vibration. But, you're only seeing half the vibration. The forward half of the forward/back.

H: Yes.

G: But why does causality go in only one direction? Why can't the effect precede the cause? Why can't all cause-effect sequences be reversed?

H: Because we're only perceiving one direction of time. Consider the cycle that you've reduced down to an oscillation. You see a point moving back and forth, back and forth, from one extreme to the other.

That's a fluctuation in a certain value of one dimension. It's entropic motion, or heat, and if conventional physics is to be trusted the universe will eventually have no motion left but that. All symbolic forms will have broken down into those simple motions. But when you consider this essential motion in the macro-context of a universe going forward in time, all you're seeing is the forth, forth, forth movement. Not any of the back, back, back. Your point will go from here to there and vanish, then re-appear here again and go to there, and vanish, etc. This is causality. We see A causing B, X causing Y, but never the other direction of the vibration. Never do we see B causing A, or Y causing X.

G: I grasp that in the abstract. I fail to see how it translates to more concrete experiences of cause and effect.

H: Well, causal sequences in real life are no different. They're aspectual views of underlying cycles. Until they're understood, they're mystifying, but as soon as you see the whole cycle, you demystify the causality. Everybody knows that sexual intercourse causes childbirth; it's a causal connection that you can cite despite ignorance of how it happens. But after biology educates us about the underlying cycles, like estrus and pregnancy, the causality is demystified. The procreative processes were shrouded in myth until not that recently, but westerners have culturally demystified them. We can even duplicate some of the natural processes in the laboratory. The mystery is gone, supplanted by understanding. I'll give you another example. When you're a child you know that the setting of the sun causes the stars to appear. You have noticed a causal link, but don't understand it. Not until someone explains to you the cyclic motion of the earth's rotation on an axis do you see the underlying cycle of that causality. Then, you realize that it has nothing to do with causality; it's merely cyclic motion.

G: But these are rather selective examples. It isn't quite so easy to see the cycles underlying more immediate causal connections. Like, why does glass break when you strike it? Or, why does uranium 238 decay into 235? Why does water flow downhill? These phenomena are documented well enough to describe with physical laws, such as the laws of motion and the laws of chemistry. But, as you've observed, describing them doesn't necessarily explain them.

H: Why does water flow downhill? It flows downhill so that it can collect in a great ocean basin, from which it can evaporate and come down again as rain to feed the mountain stream that flows downhill. You see, that's a perfectly valid explanation. Water goes downhill because our world is made up everywhere of cycles. If that cycle weren't there, we wouldn't be here. Our point-of-view wouldn't exist. As for glass breaking and uranium decaying, we can give straightforward physical explanations of them in terms of balances of dimensional values. Just as we can with water pulled downhill by gravity. But instead, let's take a larger view of what's going on when those events occur. We exist in a reality which must illustrate with its symbolic forms exactly how much time has gone by. Therefore it must steadily evolve those forms. Well, you can't make an omelette without breaking eggs, can you? Form must be vulnerable to force, even as it is drawn together by force. And there must be ideal balances of the fundamental forces. Those requirements explain, respectively, the breaking of glass, the downhill flow of water, and the instability of uranium 238.

G: I still don't see the cycles.

H: Okay. Volcanic processes melt silicon and then let it harden into quartz crystals. The quartz is eventually exposed to the elements by erosion of mountains. It is pulverized into sand. The sand is scooped up by people, melted, and formed into glass. You break the glass enough times, you have silicon dust again. A cycle. Uranium 238 is made in stars, so that's a much longer cycle, but try to comprehend this: the entropic energies that are released in the decay are themselves only aspectual illusions, just like causality is. When a photon or a neutron streaks through a bubble chamber in the physicist's laboratory, it isn't really a "free" unit of energy, though it looks so. It is one-half of a vibration, which itself is a dimensionally reduced wave or cycle. We don't see the full vibration, or the wave, or the cycle, because we are a point-of-view, and *a point-of-view can only see things from its point-of-view*. We are in a particular location at a particular time, seeing a particular symbolic reality and watching it change into a somewhat more evolved symbolic reality. The change looks like kinetic movement, but that's only because we are here, and now. If we could see everything, we would be God! We would be the universe itself! Look, have you ever been in Manhattan in mid-December? All around is frenetic activity. Seemingly random movement of cars, of people, weaving, stopping, darting, changing directions. Is any of that kinetic frenzy truly random? No. Every bit of it is part of some larger rhythm or cycle. Cabbies going around the block, tires screeching, business people returning home, hookers turning tricks, Christmas presents once again being bought, Salvation Army volunteers ringing bells, acquaintances waving to each other, snow falling—all of it rhythms and cycles. That's our universe. It's made up of an infinite number of interwoven cycles, some of which we see fully, some of which we see only as rhythms and waves, some of which we see only as vibrations or oscillations, some of which we see only as random kinetic motions.

G: And how we see it depends on how many dimensions of the cycle we perceive?

H: Of course! Even the light which makes things visible, it's a cyclic motion—going forward in space because time is passing. Sure, you can dimensionally reduce it until it appears to be a wave, and dimensionally reduce it further until it appears to be a particle, but in its fullness it's a cycle, spiralling forward. What is it really, a wave or a particle?—that's an inane question. Light is just a dimensional warp. That's what the whole universe is, an interference pattern of four dimensions expanding and warping into three.

G: I'm beginning to believe you.

H: Unirealism is understanding that, and realizing that this blueprint dictates how all change must occur.

G: And that all great questions and debates and paradoxes are aspects of this paradigm.

H: Yes. Including that of causality, which fascinated Aristotle, Bacon, Descartes, Spinoza, Kant, and every other philosopher worth reading.

G: Let me try to state the unirealism argument. All phenomena of change are very short causal sequences in much larger cycles, viewed aspectually.

H: Yes. You asked why we never see such causal sequences reversed. Well, we do, in the sense that any cycle brings a phenomenon back to its starting point. For example, why doesn't water ever flow uphill? But it does—by evaporating out of the ocean and carrying on wind currents over the mountains. And you can show similar causality-reversal with the other phenomena. It's just that time doesn't go straight forward, linearly, as we assume it does. It curves around. You recall that one of our original propositions was that time re-enters the symbolic reality by curving the dimensional warps into cycles? Time is a dimension which has both a positive and a negative value, as well. So it must go both forward and back. As with the other dimensional warps, we must be able to see it all, positive and negative value alike. But, as with the other warps, we have set up our blueprint so that we perceive the positive and negative values in two different ways. The overall effect of time going both forward and back is the creation of a cycle, and the further that time goes forward and back, the further cycles develop and the more evolved becomes the symbolic reality.

G: Now, if I'm not mistaken, conventional theory considers the forward flow of time a fundamental of thermodynamics. The arrow of time is irreversible, and entropy cannot reorganize itself into negentropy.

H: Yes, yes, and so physicists see only the crumbling of form as time goes forward. This is akin to seeing water always flowing downhill and never noticing that it evaporates and goes back uphill to make rain. And this is my point. We don't see time going backward because the symbolic reality absorbs it, by evolving. By developing cycles. Time does go backward, but in doing so it creates new form.

G: So, any event can be looked at as a cause and/or an effect of another event, but in any event what we are seeing is a short section of a cycle. However, we haven't yet discovered the cycle, and so we're perceiving the event in too few dimensions. Which is to say that our view of the event is of forward motion of a vibration or oscillation. But it doesn't vibrate back, because it isn't really occurring linearly. It circles back, because it's occurring cyclically.

H: Fair enough.

G: Whew.

H: There's an indirect approach to the realization that all causal links are only partial aspects of cyclic motion, too.

G: That being what?

H: We noted, a couple conversations back, that in our reality anything that occurs immediately begins curving into a cycle. Do you recall the roadrunner?

G: Yes. And the photon shooting through space. The conclusion was that any symbolic form must also curve cyclically. Which proved to me once and for all that I'm sooner or later to die.

H: Okay, since all phenomena assume cyclic behavior in our universe, then any event that occurs has a cycle underlying it. Provided that you take a long enough view of it from the correct aspect.

G: I see. Well, now that I understand causality, let's guess how God can be fitted into a universe where everything is causal in nature. I'd suggest this: conventional physics cannot be reconciled with the Godhead because it doesn't see beyond causality sequences, but you must see beyond causality to explain the emergence of form in our universe. If there were nothing operating but strict causality, there would be no galaxies, or solar systems, or human beings.

H: I would agree. Causal sequences are linear. In unireality, if you follow a long string of causal links you see that the string itself becomes a cycle, or conjoins with other causal strings to form cyclic behavior. Recall our photon travelling unimpeded through the universe? Gravitational fields causally curve it here, then there, and the net effect of all these causal nudges is to shape the trajectory into an incalculably large circle. Metaphorically, this is like getting lost in very deep forest, where you can't orient yourself to distant mountains. If you set off walking, you will wander in a very large circle, because it's inevitable that you'll slightly favor one side or the other. Even though you conscientiously set one foot directly in front of the other. Causal links, no matter how deliberately or randomly assembled, end up making patterns, and patterns are always either literally or figuratively interference patterns. Patterns are made up of rhythms and vibrations. How does a child develop a pattern of speech, out of babble? By subtle reinforcements and corrections during years of interaction with older siblings and adults. The net effect of all these causal influences is the emergence of a pattern of speech. These examples are not isolated. In any sphere of reality a series of causal influences ultimately shapes into cyclic behaviors in space-time. And such cycles compose the interference pattern of a symbolic reality.

G: Which is why we have form at all.

H: Yes. A point-of-view is an interference pattern of background cycles. Now, it's possible to analyze any such interference pattern into parts that seem wholly unrelated except by causality. Our universe itself, if you look at it analytically, is nothing but a great number of randomly-located oscillations. But as a whole, the whole is far greater than the sum of the parts. And, if you look at the problem this way, the whole is there because the point-of-view is there; the whole is there because our point-of-view integrates all those oscillations into form. Which is paradoxical, is it not? Without the interference pattern of form, there would be no point-of-view, but at the same time form exists only insofar as we see it.

G: As an analogy, I can reflect on my personal history as a meaning-less jumble of events. But if I do, I cease to have a sense of continu-ity as a person. So what I am is the perception of sense in those events.

H: And, you can look at the universe as a jumble of unrelated vibra-tory motions, but then you can't see the form, which means you can't even exist, as a point-of-view. If the universe is nothing but heat, you don't exist. Similarly, you can look at the universe as if it is nothing but causal sequences. In other words, as vibratory motions in one di-rection of time. But so what? It's reassuring to know that all events occur in causal sequences, but the fact falls far short of telling you what's going on. And the rest of what's going on is God-like. That's where God is. He's to be found in the creation of form.

G: The myth of Genesis.

H: God created form out of Nothingness, and He continues to do so. But He is so wrapped up in His work that you cannot extract Him from it to set Him on a throne. He isn't some anthropomorphized de-ity. He is the process of sense arising from the essential dimensional warps. Things change, and He is there in the way they change. Things remain the same, and He is there in the way they cohere.

G: Causal sequences, because they are linear, do not reveal the pres-ence of the Godhead.

H: Yes. The Godhead is in the form itself. It's in the full cycles, and you won't notice it if you only pay attention to the collapsed cycles— to causal connections. We discussed the cycle of precipitation. Wa-ter flowing downhill to evaporate from the ocean and make rainfall later. Every step along this cycle is governed entirely by causality. Water flows downhill for causal reasons. It evaporates for causal rea-sons. It forms droplets for causal reasons. And it falls for causal rea-sons. Big deal, right? You can't impress an atheist with those steps. What's amazing, what sings of divine presence, is that the design of this cycle constantly replenishes the supply of clean fresh water on the continents of planet Earth, so that land is inhabitable. And it's that aspect which persuades people to believe in God. A physicist gazes at the stars and sees causal sequences. Then he goes home and gazes at his newborn child and sees God. He becomes convinced that only a divine creator could have designed all these many causal links in such a way that the world could bring forth life. Well, of course there isn't really a divine creator who monkeyed around with ineluctable steps of causality until he'd come up with an interactive background of connections that could sustain life-forms. Anyone who believes that is assuming that causal connections are logically prior to form itself. They aren't. Listen to me, they aren't. If it weren't for the form, there would be no causal connections. Form is logically prior to causality, because you can't have a universe until you have a point-of-view, and you can't have a point-of-view without form, and you can't have form without all those underlying cycles. Causality doesn't enter into this picture until you analyze, and *find*, causality.

G: Which must have happened early in history. And that implies...

H: That God has been in retreat since the very origins of human society. The moment you begin analyzing events into causal connections, you simultaneously invent God and begin driving him away. In the beginning, when consciousness is only nascent, God is an unnamed force or spirit that imbues the world. This is important to understand. Human beings invented God—first as spirits, then as deities, then as one grand deity—as consciousness emerged and then evolved through stages. Thus the god-concept is a product of consciousness. For the first few millenia that concept, the idea that there was a divine creation and was still a continuous manipulation, was the glue that held together a host of causal connections. But where causality is, God isn't. And as the causality-ruled sphere expands, the divinely-ruled sphere shrinks. It shrinks, and it sharpens into focus. Eventually, all of physical and mental reality is explained by causal connections, and then all you require of God is that once upon a time He created it all, set things in motion. Hence Newton's version of God. But it was the big bang theory that finally got rid of God entirely, as you have pointed out. The big bang doesn't need a creator. It needs only random chance to occur.

----Ten----

Grana: Physics explains reality in terms of laws. Unirealism explains reality in terms of symbolic representations of Nothingness. Will it be possible to reconcile these two different explanations?

Heidrich: I believe so. Those laws of physics were discovered empirically, with a dash of intuitive guesswork here and there. Centuries of observations give them validation, but that's about all. They are obviously true, but no one knows why they should be. It isn't brash to suggest that there is a deep structure to reality that fits all these laws together and thereby indicates why they are true.

G: Newton's laws of motion, Maxwell's equations for electromagnetism, Einstein's equations for relativity, and quantum mechanics as well?

H: Yes. If, as I argue, the universe is zero-sum, manifested through expanding dimensions and symbolizing, then all things and events within it can be interpreted in terms of symbolic transformations. Until now we have assumed that things exist and move about in a dimensional matrix. My counterproposal is that we have a dimensional matrix and nothing else. The form we perceive is a 4/3 representation of the overall value of those expanding dimensions.

G: And how do we get such a symbolic reality from Nothingness?

H: The Nothingness is separating into four binaries, each of which has a positive and negative value. By using one of the binaries to link together what you have arbitrarily designated to be the positive ends of the other three binaries, you begin creating a symbolic reality. The negative ends of the binaries twist around and meet the positive ends to form balances.

G: They are bending into the 4/3 representation.

H: Yes.

G: For example, if I have a cube made of pipe-cleaners, and I flatten it down to two dimensions, the wires will bend.

H: That's a crude analogy. What's going on is far more interesting. It appears that each of the binaries is twisting like a Mobius strip does, with the ends meeting, but in reversed positions. So that in our physical reality we perceive both ends of the binary, but the ends are mirror images.

G: The Mobius strip is a two-dimensional strip of paper, twisted around with its ends glued together.

H: Yes. But by putting the twist in there you're adding a new dimension. The two-dimensional strip of paper now curves through three dimensions. It's an especially good analogy because it resonates with the concept of *spin* in subatomic particles. The odd feature of the Mobius strip is that if you trace a line on it, you must go around twice to get back to your beginning point. Spin also goes around twice to return to its original state.

G: So the binary starts out with no dimensions. The twisting is what's responsible for making dimensions.

H: That's a significant observation. Each binary must originally be a dimensionless point. I don't see how we can talk about dimensions existing prior to the creation of form. And I think it is clear that the twisting, whatever that means, is what invents dimensions. We're accustomed to thinking that three-dimensional space just exists, independent of what occurs in it. It's more likely that space is an illusion created by the twisting, and that the only reason there are four noticeable dimensions is because we perceive binaries twisting four ways.

It is even likely that one binary is twisted through one dimension, a second through two dimensions, a third through three, and a fourth through four. At any rate, those four binaries provide us with a three dimensional space, because the linkage constructs a 4/3 representation. The fourth binary manifests as time.

G: So our universe is constructed of binary sets, which are twisted with the ends reconnected but reversed. And our universe has lots of these twistings.

H: From here on I'd like to call them *warps*.

G: Our physical universe is composed of warps, and what else?

H: That's it. Physicists will tell you our universe is made up of matter and energy. Well, it's actually made up of dimensional warps. You see the warps where the binaries are twisted, and you see the positive and negative ends. That's all. There are four types of warp. We call them the strong force, the weak force, the electromagnetic force, and gravity. We call them forces because they appear to force particles to move in certain ways. But they're warps of space and time, and a particle moves in these warp-fields just as a child goes down a slide at the park. Even that analogy is misleading, because the particle itself is made up of dimensional warps. The four different types of warp are heterogeneous, in that the gradient of the warps differ greatly, and they also interact in distinctive ways. The gravitational warp, for example, affects all other warps, but only slightly, whereas the strong warp affects only particles called baryons, which themselves are dimensional complexes of three or four kinds of warp. I haven't figured out exactly why these warps interact in the way they do, but I state with absolute certitude that they are what our universe is made up of, and all that our universe is made up of.

G: And the warps have a binary nature, so that each of them zero-sums?

H: Yes. So if you broke the linkages somehow, the universe would vanish, in theory.

G: I presume that the warps manifest as fields. What about the mirror-image ends?

H: Those we see as particle characteristics. Charge, spin, mass. Everyone is familiar with the positive and negative poles of electromagnetism, of course. And the weak force is strongly associated with the neutrino, which is a massless particle closely resembling a one-dimensional Mobius strip, and which has two forms—one spinning clockwise, the other counterclockwise. The gravitational warp also has two components, the one we are all acquainted with pulling everything together, but the other, with which we are not so well acquainted, pulling everything apart. That latter component is what causes the expansion of the universe, and it isn't unfair to call it anti-gravity. I know it is hard to visualize, but the universe is gravitationally warping both ways at once, kind of like turning a sweater inside out. As for the strong force, it has an active range that's barely the width of a proton, so physicists know far less about it than gravity and electromagnetism, which both have infinite range. Given that short a range, it's not surprising that a countercomponent might not be easily discerned. Experiments have tentatively measured what looks like a fairly short range repulsive force that acts on hadrons, though, and I'd consider that a possible candidate.

G: So the universe is expanding for the same reason as it is being pulled together? The gravitational warp?

H: Yes. Because of the design of the linkage, the negative end of the binary used to lock together the positive ends of the other three binaries becomes an anti-gravity that steadily increases the dimensions of the continuum. The pulling together and the expansion are complementary. Physicists don't yet realize that gravity and the expansion—which I call anti-gravity—are two sides of the same coin. Equal opposites.

G: Well, it's not readily apparent how you can have a slide in the park that goes up simultaneously as it goes down. You seem to be saying that every gravitational field is overlapped by an antigravitational field, which defies both common sense and common experience.

H: On the contrary. Common sense and common experience both inform us that nothing stays in place unless is it held in balance, and what force counteracts gravity except anti-gravity? The warps each have a vector. In the case of anti-gravity, the vector is outward. In the case of gravity itself, the vector is inward. Unlike the magnetic field though, which affects only charged particles, or the strong force field, which affects only baryons, or the weak force field, which is also very particular, gravity and antigravity apparently affect everything. Everything slides along the gravitational warp—slides together and apart at the same time. And the only way you can can visualize that is in the context of four full dimensions, and none of us are trained to think that way.

G: I'm certainly not.

H: Nor am I. Yet it is happening, and it's a terrible puzzle even for physicists, because it's evident to them that the more things fly apart, the more gravitational potential energy they have. If you carry a brick to the top of a ten story building, with every step you climb the gravitational potential energy of the brick increases. But in that instance, you are providing the energy, you're doing the work. In the case of the universal expansion, the increased potential energy of all those galaxies flying apart is coming from nowhere. You can't say it comes from the big bang itself without raising serious questions about what energy really is. After all, before the big bang is presumably Nothingness. Since when does Nothingness explode? Where does the force of the explosion come from? Unirealism explains this conundrum in terms of zero-sum binaries, but conventional physics is at a loss, and deliberately ignores the question.

G: Conventional physics, you said earlier, describes the universe in terms of matter and energy.

H: As if that's all there is. But if that's all there is, what is a charge? What is a spin? These qualities of physical reality simply don't translate into matter or energy. You can't turn a positive charge into either matter or energy, and you can't manufacture a positive charge out of them, either. The exception is, of course, manufacture of pairs of matter/anti-matter particles, and those self-cancel back into energy. So, then, what is a charge, or a unit of spin, if they can't be translated into matter or energy? Obviously, then, you can't describe the universe exclusively in terms of matter and energy. Whereas you can explain it exclusively in terms of warped dimensions made from zero-sum binaries.

G: And matter consists of balances and counterbalances of warp-fields.

H: Yes, with protons and electrons being the quintessential balances—the linkages. The warps are deeply, deeply interwoven, in the tapestry that is physical reality. Warps are fields, with vectors—or, to be more accurate, two equal and opposite vectors, depending on which end of the dimensional warp you specify. But if you overlap and intersect these fields, you can make a particle appear momentarily, before it fades back into its field manifestation. Okay?

G: Like a whitecap appearing at the peak of an ocean wave.

H: Very much like that. And all this tapestry of interwoven warp-fields is an interference pattern of the underlying cycles of balanced warp-vectors. So that when you see a change in the interference pattern, you're seeing the movement of standing waves. If you think of a proton as a node in the interference pattern, then when the proton moves the node is moving. It isn't motion at the speed of light, which is how fast the underlying warps move, because it is the movement of a standing wave. And the interference pattern is constantly roiling with change. Time is passing, and being steadily integrated into the pattern, and the symbolic form is evolving.

G: Whenever I move, or move something, I'm re-arranging the interference pattern.

H: Yes. You may say you have used energy. All you have really done is alter the underlying arrangement of warps. If you jump to the conclusion that the warps *are* energy, you're missing the point. It is change in the *pattern* of warps that is energy. Like the child on the slide: the slide itself is not energy, it's the motion of the child going down the slide that is energy. The slide is merely a field with a vector. When you re-arrange the underlying warps, you change the vector situation, and things move so as to accord with what the new vectors are. All energetic processes are characterized by imbalanced vectors. The vectors may be a random jumble, as in an explosion. It is even accurate to say that energy is nothing but vector quantities. And the vectors come from the vectors of the warps.

G: So the interaction of warps, that looks like what we call energy?

H: Yes. A gravitational field is not energy, it is a dimensional warp. Now, if you drop a particle into that field, it will move in the vector direction of the warp, which is to say toward the mass at the center of the gravitational field. The particle will fall. As it falls, its kinetic energy grows, hence its relativistic mass increases, and the sharper the gradient of the warp, the more kinetic energy the particle takes on. Remember now, the particle itself is made up of warps. So energy is the activity of warps affecting each other. Energy is an indication of how the dimensional warps are interacting.

G: And potential energy is equal to the warp-vectors in balance?

H: Sure. Even if you profoundly disturb the interference pattern, it gradually settles into a new pattern. Electron fields and neutrino fields and proton fields assemble themselves into stable complexes we call atoms, atoms bond into molecules, and then there's the ever-present gravitational warp, mildly insistent that all other warps weave into the universal tapestry. When warps are in these counterbalanced complexes, the arrangement of patterns has what we call potential energy. Once the warps are moving each other about, the energy is called kinetic. In the former case you have *things*, in the latter *change*. So any symbolic reality contains enormous potential energy.

G: Which can be released?

H: In which case it becomes kinetic. Disturb the warp balances, and everything starts moving, much of it at the speed of light. The symbolic form breaks down. At least from your point-of-view. Because symbolic form only exists in symbolic reality, which depends on a point-of-view. That's why physics uses the term frame of reference—to establish point-of-view. So, kinetic energy is in motion relative to a given frame of reference, and potential is in stasis relative to a given frame of reference. The distinction is more than just semantic. According to the equations of relativity, a symbolic form that's in motion has more mass than a symbolic form that's standing still. It has relativistic mass in addition to its rest mass, which is to say it has kinetic energy. But since there's no universally objective measurement of what's in motion and what's standing still, the perception of relativistic mass in a symbolic form depends entirely on whether or not you're moving in relation to it. Anything that is standing still according to your frame of reference has only rest mass—i.e. potential energy—and no relativistic mass—i.e. no kinetic energy. So time, distance, mass/energy—and let's not forget magnetic fields as well—all depend on point-of-view.

G: Which explains why you prefer, in unirealism, to talk about symbolic reality, rather than reality.

H: Yes, because reality is all in how you look at the zero-sum Nothingness. Reality is always composed of symbolic form, but for different observers the form looks different. I'll give you an easy example. We can look into the night sky and see a faraway quasar exploding. But to an alien who inhabits that section of sky, the quasar exploded thousands of years ago. Thus, we see different form in the universe—which is a symbolic reality—because we have different points-of-view.

G: Of course, even the standard model of cosmological physics accepts that.

H: Accepts, but can't explain, because the model still can't escape the prison of objective reality. To give an indication of what unirealism is up against, listen to this quote from Albert Einstein himself. "The belief in an external world independent of the perceiving subject is the basis of all natural science." Einstein accepted relativism of the features of reality, but couldn't bring himself to accept relativism of reality itself. But once you accept the relativism of reality itself, understand that it's an evolving interference pattern of dimensional warps, as viewed from a particular point at a particular time, the problems and paradoxes of science vanish. You can explain things that otherwise you have to accept on faith alone. Inertia, to name one. Why should a body in motion continue in motion—but only from point-of-view of the frame that accelerated it? Why should a body stay in motion at all, unless you have altered the warp-vectors of the interference pattern? If you have a standing wave in a pattern and you change the constituent waves slightly, the nodes will begin to slowly move. They will continue to move as long as you don't re-balance the constituent waves. Well, there's your body with inertia. It's just one of those nodes in the interference pattern. You see?

G: But eventually this physical reality of ours will run out of energy, won't it? If things get further and further apart, so there is less inter-action...

H: Okay, cosmologists operate under certain assumptions based on laws derived from observations. Given those assumptions, the universe will either expand indefinitely into total entropy—heat-death, you might call it—or it will gradually slow down its expansion until gravity gets the upper hand and pulls everything back into a singular point again. Which scenario plays out depends on how much mass there is in the universe. Is there enough gravitational pull to close the universe, or not? But, assumptions are assumptions, no matter how self-evident they may seem to be. And I suspect that in this case the assumptions of cosmologists are wrong.

G: In fact, you seem to be suggesting that as the space-time dimensions of the universe grow, so does the amount of energy in the universe.

H: All right, here's how it seems to me. The expansion of the universe steadily increases the amount of warping. Not the gradient of the warps, but rather the extent of the warps. If this is true, then the amount of interaction occurring as warps juxtapose and interact also steadily increases. Hence, since we define energy as the interaction of warps, the amount of energy in the universe must also increase. And, of course, mass. The universe, then, gains total mass as it expands.

G: And the increasing interaction of warps manifests as increasing complexity of form.

H: We might not discern a gradual increase in total energy. Let me explain with an analogy. The most distant galaxies we can distinguish are characterized by red-shifts suggesting recessional velocities of greater than 90% the speed of light, but that rate of expansion is a function of the distance from which it is observed. We don't see nearby galaxies receding that fast, and neither would we see dramatic local increases in energy or mass. It may or may not be significant that we *do* see dramatic outpourings of energy in distant galaxies, I don't know. It may be that the value of rest mass is a function of the expansion, spatially and temporally, of the universe, in which case that value is imperceptibly rising.

G: Fred Hoyle, the famous astrophysicist, used to promote the idea of a steady-state universe in which matter spontaneously appeared, in miniscule amounts. I gather the big bang theorists finally refuted him.

H: Unirealism validates both points-of-view, to some extent. Yes, there was a point of origin, when expansion began, and yes, there is evidence, such as the background radiation, which indicates that to be true. But yes, the universe is gaining mass, and yes, it is an eternal universe. By the way, Hoyle was less famous than infamous. He's the one who invented the term "big bang," and he used the term mockingly.

G: Still and all, what you've said isn't tantamount to heresy.

H: I reason further, though. Consider something more. If the universe's expansion stops, then time itself must stop. After all, time is just another expanding dimension. If time stops, how can we still have vibrations, waves, and other energy exchanges in the universe? Because without time, motion has no meaning. So I must conclude that without the universe expanding, there would be no such thing as energy. All energetic processes would come to a grinding halt. I realize that's an incredible assertion, but the logic of unirealism suggests strongly that it is the universal expansion that drives everything around us. When a photon escapes from an atom, why does it fly off at the speed of light? Why doesn't it just sit there? Why does a neutrino streak along at the speed of light? Why doesn't it just sit there and spin? Isn't it possible that their vector motion is a direct expression of the universal expansion? I suspect that the vibrations of the subatomic particles, the shooting photons and neutrinos, that is all froth created by a constant infusion of dimensional values. The dimensions of the universe grow, increasing the extent of the warping, but the warping is not out there somewhere—it is everywhere, in all the tiniest regions of space. And since the warps are all entangled, the woven tapestry of the interference pattern is ever reorganizing itself. The vibrations and waves induced by the expansion gradually integrate into new form.

G: We can safely call that heresy, I think.

H: Scientists have always taken for granted that those elemental vibrations and rhythms and waves are somehow just there, intrinsic to matter. But, why? It's not so farfetched a proposition, what unirealism says. We have a universe that constantly expands, time that constantly goes forward, and constant motion within that universe. Why shouldn't these phenomena be connected logically, in an elegant unity? I find current theory—not just cosmology but most of physics—a confusing amalgamation of *ad hoc* propositions. Unirealism may not yet be proven true, but there's a wonderful elegance in it. And I like to think the universe itself is very elegant.

G: So, you aren't dispensing with what is known by science to be true. You are putting it into a larger context, so that we can discern the full relationship of the parts. The essential point that unirealism makes is that all of this springs out of a zero-sum Nothingness. Which is as yet an unproved hypothesis in cosmology.

H: Still, you notice the evidence of the zero-sum all around us in everyday life. If you take a fresh look at phenomena like homeostasis, bifurcation, and bilateral symmetry, doesn't it strike you that our world is nothing more than expanding balances? Our world isn't deteriorating; it's opening up. Blossoming, like a rose. Growing outward, like a fern. Some balances we see, some we don't, but the incredible order of our universe should inform us that this isn't just some meaningless jumble. So, when we document one of the aforementioned laws of physics—such as "every action produces an equal and opposite reaction"—we might do more than treat it as an isolated rule. We might ask why there should be such laws at all. And ask ourselves this: isn't it possible, isn't it in fact probable, that all such laws fit together in a single structure of transformation? And, wouldn't that new science of *things* and *change* explain everything that exists, including our perception of it? And, most importantly, isn't that new science a satisfactory explanation of how Somethingness emerges out of Nothingness?

----Eleven----

Grana: If reality is indeed subjective, and if each of us is a point-of-view, then why do all of us experience the same universe?

Heidrich: Strictly speaking, there isn't a universe. There is only perception of a universe. The form that you perceive—be it stars and galaxies, trees and mountains, or your own body—is there as part of an interference pattern that describes this moment in time, which is the moment in time at which you, as a form, exist. But the form is a fantastic illusion, outside of the point-of-view. Only to the point-of-view is it real. So, the universe is commensurate with you. The universe emanates from you.

G: And from you. And, from her, from him, from them.

H: The fact that the universe is commensurate with you, and the fact that it contains other sentient beings with whom it is also commensurate, are not contradictory. Obviously, if you were not all experiencing identical universes you wouldn't even be aware of each other.

G: So, the apparently objective background we all see is a single universe, but single to each one of us as well? All our individual universes, when overlaid, coincide perfectly?

H: Yes. And presumably if an alien showed up on our planet, he would also see the same universe as you and I do. Otherwise he couldn't show up.

G: But this isn't the same thing as an objective universe?

H: Correct. Here we are dealing with two paradigms. The conventional one is that we are forms moving through an objectively real universe. The unirealism paradigm proposes that we are points-of-view that are not mutually exclusive. Under most common circumstances, the two paradigms provide identical experiences. Both allow a shared reality. Not until you examine very closely how change occurs in this shared reality is it evident that the two paradigms are not synonomous. Today, technology permits us close scrutiny of ongoing events. Both macrocosmic examination, with telescopes, and microcosmic examination, with particle accelerators, indicate that the first paradigm doesn't give the right results. The new paradigm does. Which is why physicists like Heisenberg and Wheeler and von Neumann rather reluctantly began to suggest a many-worlds hypothesis. That hypothesis suggests that we see only one universe of an infinite number of universes, and that we generate new universes any time we make an observation. Philosophically, however, if the others do not exist because we cannot see them, then this one exists only because we can see it. It's not that there are many universes, but that there are none at all. There's only an illusion of a universe, conditional on a point-of-view implicit in that universe. So in my view, the many-worlds hypothesis is a desperate attempt to salvage the conventional paradigm. It doesn't go far enough. It still assumes that the universe has objective reality. The many-worlds hypothesis does reconcile some theoretical problems, such a qwf collapse, with conventional reality. But unirealism resolves all theoretical difficulties, I think.

G: Such as?

H: The EPR paradox, for one. For that, we must be familiar with the concept of *correlation*, a phenomenon that subatomic particles exhibit. It goes like this: when two or more particles interact, they become correlated. The immediate consequence of the interaction is that they lose their distinct identities and assume an undifferentiated common identity. Whereupon you have the original particles, and a set of characteristics, but those characteristics are not specifically assigned to either of the two particles. In effect they share them. However, if you make an observation on one of the particles, it will instantly have a proper set of characteristics, and also instantly the other, unobserved particle will have the remaining characteristics. In a very real sense neither particle has specific characteristics until the moment of observation. This sounds odd, but not especially impossible. Until you consider the implications, which is what Einstein—for whom the E in EPR stands for—did do. He suggested an experiment by which you allow two particles to interact, and then send one of them off into the cosmos at the speed of light. After a while, you make an observation on the second particle, which you have trapped in your laboratory. As soon as you make that observation, the particle takes on a specific identity. But, simultaneously, the other particle— which by now is halfway across the galaxy—will assume the leftover identity. Now, it's a random choice which identity that second particle, the one in the lab, will assume. So, how does the first particle, so far away, know which one it must assume? There simply isn't any way for the information to reach it fast enough—unless the information travels at superluminal speed. To Einstein, this situation wasn't just peculiar, it was impossible. Quantum physics was either wrong or incomplete. The paradigm was flawed. Years later, a mathematician named John Bell proved beyond any doubt that the thought-experiment of Einstein's was an accurate depiction of how the world worked. In other words, reality is not local. It is non-local. Events happening here and across the universe are connected instantly.

G: Which one would assume is impossible in an objectively real universe. Because you couldn't cause a distant event to happen any faster than a light wave could arrive there to trigger it.

H: Whereas, if the universe has no objective reality, if it exists strictly and solely in point-of-view, then a change in the point-of-view would radiate instantaneously through the universe, provided that it didn't involve an increase in the total dimensional values of the universe. And, that criterion is met by the observations in the EPR experiment. No new dimensional values are being introduced. The current ones are not even being exchanged or modified. In fact, since the phenomenon doesn't involve any evolution of cycles, it *cannot* take time. Time, you recall, is one of the expanding dimensions, and its passage is steadily integrated into the cyclic structures of the interference pattern. So time does not pass unless the structures change.

G: It's as if the universe is inside my head, although I imagine it to be out there. So, at least some of the changes I make in how I imagine it will occur instantly.

H: Yes. Changes in your thought that involve reasoning will take time, because you're moving things around. Assembling them in this order, trying them in that order, until you find an arrangement or sequence you like. But changes which only involve a change in perspective can happen in a flash of insight, a sudden understanding.

G: And the EPR experiment, which presents a paradox in the conventional paradigm, is quite acceptable in the paradigm of unirealism.

H: Also, this paradigm of subjective reality illuminates other problems. For example, the apparent existence of telepathy. Because telepathy is very similar to the communication between particles in the EPR experiment.

G: But, certainly subatomic particles can't communicate telepathically. They have no brains, or language…

H: But they interact, and so become a single identity, and no matter how you spread that identity out over the cosmos, it remains a single identity. I'm suggesting that the same thing happens when human beings interact, although with human beings the dynamics of the correlation are much more complicated. These two particles only share a few attributes of identity, whereas a human being possesses an incalculable number of attributes, and in that complex of attributes observations are occurring all the time. In the particle world that isn't happening, so two such particles can remain correlated indefinitely. Nevertheless, such correlation between human beings does occur, as a result of interaction. You're married. You are probably aware of a certain sharing of identities between yourself and your wife.

G: Often I catch myself thinking as she would. And there's a sort of silent communication between us that verges on telepathy.

H: When two people interact and form an emotional bond, they become correlated like that. So well correlated that they acquire a virtual psychic connection. The connection defies the known laws of physics. Especially when it traverses great distances instantaneously. Of course, there's no way to test whether telepathic awareness exceeds the speed of light, but I am confident that it does. Now, a curious thing about telepathic messages seems to be that they usually signal a change in identity. In other words, when a psychic connection exists, it's noticed only when one of the partners of the connection strengthens or weakens the correlation by engaging or disengaging the shared identity. Otherwise the correlation goes unnoticed. And isn't this like the particles, with their telepathic messages? They only send a message when the correlation is broken—which is to say when the shared identity is disengaged.

G: But that correlation is broken from outside. By the experimenter who makes the observation by measuring something.

H: True. A correlation, being a shared identity, cannot break by itself. And, when two people are correlated, what is the outside agent that breaks the correlation and restores distinctly separate identities? Why, the consciousness, which rises above the self and surveys what's going on, by means of cognitive symbols. The consciousness is performing the observation. What unirealism proposes is that under normal conditions the members of a society, because of their interaction, enter into a correlated state that is analogous to the subatomic type of correlation. They begin to overlap identities. Then, rather than having an amalgam of individuals, you have a tribal identity that subsumes the individuals.

G: And to the degree this occurs, there is a single shared universe.

H: From a single shared point-of-view. So, the belief that there is a single shared objective reality is correct, up to a point. The reality can become so deeply shared that it might as well exist objectively. But *might as well* isn't the same as *does*. Observations will still interrupt the correlation. The extent to which that happens depends less on the number of observations occurring than it does on how radical such observations are. As a point-of-view, you're continuously making observations, you're continuously collapsing qwf's, even when you're not thinking about it. By breathing, by sensing, by the beating of your heart, you're observing. By eating and digesting, by copulating, you're advancing your frame of reference into the future. Every time a photon enters your eyes and registers there on the retina, you've made an observation. Every bit of food you break down and release energy from, every sound you hear. You can't stop observing. If you do your point-of-view rapidly disintegrates. Such routine observations are natural to your being, and making them isn't going to have the effect of breaking the transcendent correlation you've established with a fellow human being.

G: While conscious observations might.

H: Though it is important to recognize that continued interaction can restore the correlation. Granted, during the breaking of the correlation, in the period of time where the identities are distinct, the identities may change in such a way that when correlation is restored the shared identity will be different than it was before the breaking. But correlation does tend to return if interactions resume.

G: And these dynamics of correlation and observation apply to human relations in reality that is entirely subjective. So, correlation is the perfect overlay of one subjective universe on another, and observation is the separation of them.

H: It might even be specified that an observation that separates universes is by definition a conscious one. This requires a somewhat more flexible definition of *conscious* than we're accustomed to. Let us suggest that a conscious observation is one that symbolically synthesizes various previous observations. A conscious observation, then, alters the composite symbols of the point-of-view.

G: While routine observations do not?

H: Yes. But, a series of routine observations may lead up to a conscious observation that does rupture the shared reality. For example, in our earlier discussion of the ecosystem we noted that it constituted a symbolic reality. Well, you can say that it makes conscious observations. In fact, the point of calling it Gaia is to impute to it the power of thought. Rather than using linguistic symbols for thinking, as we humans do, Gaia uses the symbols of genes and chromosomes. Gaia manipulates these as she reasons her way into the future. A mutation here, a gene transfer there. She is artfully working out ways to keep her frame flat. If she doesn't succeed, her cycles will become distorted and ultimately she'll crumble and collapse. So, when it's necessary she makes decisive changes, which are conscious observations. But, millions of routine observations lead up to such a conscious observation. The symbolic changes in the peacock's tail weren't casual or random. They were preceded by a gradual accretion of single observations, which finally were synthesized into the new symbols. As we've noted before, this isn't any different than the way you or I think. We notice something, we notice it again, and again, and suddenly we meld all these observations together and come up with symbolic information. Our unconscious observations can eventually lead to conscious observation.

G: So, whenever you have a reality shared by different points-of-view, each point-of-view is making routine observations, which sooner or later lead up to a change in the symbolic content of the reality. But under what you call normal conditions, there's a fairly continuous correlated state. By normal conditions, you mean something on the order of tribal life.

H: Yes. Or the ecosystem itself, without human consciousness interfering. The world of plants and animals is a correlated state—a milieu of integrated cycles in a frame that stays flat against time. Each of the points-of-view in it is making observations, but it stays correlated.

G: Because the observations aren't conscious.

H: The consciousness that's active belongs to the whole ecosystem. The individual members do not have individualized consciousnesses.

If you follow the progressive sophistication of life-forms, somewhere around the mammalian level you begin to see tribal consciousness emerging, culminating with the primates and especially Homo sapiens. The most primitive human tribes have a collective conscious identity, which doesn't cause them problems as long as the tribe remains isolated. But when you find two or more tribes contingent, so that their identities conflict, you also see breaks in correlation. That's when the real problems start. Either tribe by itself is balanced and egalitarian, and the members, being subsumed by the larger, shared identity of the tribe are correlated. Even if a member breaks the correlation by conscious observation, which undoubtedly happens, he afterward gets pulled back into correlation. He is obliged by his peers to undo his wrong, and once he has, the correlational condition returns. Needless to say, this is what happens in the love relationship. Falling in love is falling in correlation. Almost immediately there's a sense of shared identity, of likeness and dovetailing, and the two become one. And then the individual conscious identities try to merge, and that is not easily accomplished. The emotional bond must be extremely powerful to allow it, because it must override all conscious recognitions of irreconcilable differences. While the merging of conscious identities is taking place, there will be slights perceived, small betrayals, lovers' quarrels. If both partners are governed primarily by the desire to keep their personal frames flat, then after each of the conflicts the correlation can return. Apologies are made, commitment reaffirmed. Which is to say, everything's fine as long as self takes precedence over ego, for each partner. And that won't happen unless the emotional bond is awfully strong. Generally, you find such bonds among family members: mother-child, twin-twin, close siblings, and mates. And it is with such relationships that you find the most frequent incidence of telepathy. When the relationship is bonded with blood, you have the most durable correlation, which explains why primitive culture makes such a big deal about kinships. As for the romantic bond, it almost always loses out to the pressure of conscious thought. The romantic love we prize so highly, though, has far less significance in a society with a communal identity and a flat frame of reference. We idealize the romantic bond, probably, because of the fact that our consciousness is so extremely individualized. And that is a rather recent development in history.

G: This egalitarian tribal society you keep referring to, is there such a thing?

H: Most of them are gone now, existing only in anthropology texts. Once upon a time, long before the political state emerged, all human society was tribal and egalitarian. Ever since, society has slowly and fitfully evolved, often because of contact with political states—by which I mean societies like Egypt and Rome and the Inca and the Chinese, as well as modern states like the USA and the USSR. Yet you can find a few isolated tribes even today, tribes in the Amazon and New Guinea. There are real virtues in such societies. Tribal members have a deep sense of belonging, of fairness, of sharing, of empathy and affection and nurturing, as well as a magnificent sense of playfulness.

G: They're also stagnant. There's no progress.

H: They're static, perhaps. Not stagnant. They could join the main-stream of progress if they wanted to. They don't. They think we're insane. And maybe they're right.

G: Nevertheless, if evolution and continuous differentiation is a ten-dency in reality, it is indisputable that the tribal societies are resisting natural changes.

H: There are two distinct symbolic realities at issue here. The first is physical reality, the second is abstract symbolic reality, which is a cognitive reality of ideas and identities, books and sciences, religions and technologies. Any human society is bound to have both, because the human mind is so powerful that it's going to develop an abstract mental model that describes the first, the physical, reality. Both of these symbolic realities are guided in change by the symbolic trans-formations. The immediate difference is that the former changes very slowly and the latter changes very rapidly. Certainly you can't resist change in the former. How are you going to prevent genetic evolu-tion? Or the gradual burning up of the sun? Or the cooling of the earth's core? You can't. But you can resist development in the abstract symbolic reality. The primitive tribes do it. They defuse the developmental tendency. They protect the correlated relations by means of rituals. Those rituals are aimed particularly at preventing disparity in genetic advantage from translating into disparity in material advantage. The rituals work to spread out the wealth, so that no member gets too much control over other members.

G: For example?

H: The potlatch ceremonies of the American Indians, where goods are destroyed. The obligatory division and sharing of hunted game. Disaccumulation rituals. Traditions that require the richest man in the tribe to host the biggest banquets, and to give most generously at births and weddings. There's a pattern here: the prevention of accumulation of wealth. Also, in such tribes there's some active concept of *usufruct*, a principle that considers ownership a relative thing. If one man has a knife he's not using, and another man needs a knife, then the second man has the right to use the one that's available. For the same reason, nobody can own land. If you're not actively using a piece of land, it's open for use by anyone.

G: And these rituals and beliefs stem development of abstract symbolic reality?

H: I strongly suspect that when such material symbols of wealth are regularly redistributed, the evolution of ideas halts. The progress of ideas is driven by two forces: greed for power and security, and the resentment of inequality. As long as society is kept egalitarian, the motivation for developing new abstract symbols is limited.

G: But if the rituals fail, then abstract symbols begin to evolve? Is that why we see a general progress in human history?

H: Oh, yes. Human history is the record of development in abstract symbols. Ideas develop, and consciousness develops alongside them. Any symbolic reality has its own natural momentum, and cognitive realities are no exception. It's natural for human beings to generate new symbols. Symbols enhance effectiveness.

G: All right. In summary, when two or more people whose points-of-view of the universe are compatible do interact, their universes effectively become a single shared universe. In it, their identities become to some degree shared. And that is a condition of correlation.

H: Identity in the sense of self, because that's what correlates, not ego identity.

G: Okay. And, this condition of correlation is ideal. It has intrinsic value. True?

H: True, because when there's correlation the relationship is flat against time. The interactivity becomes a tapestry, an interference pattern. In the case of a tribal society, the cycles remain balanced. The rate of procreation sustains a fairly constant population; foodstuffs are equitably distributed; maturation from infancy to adulthood is progressive and well-defined. All in all, the tribe becomes a single organism, functioning with a flat frame of reference. A similar correlation operates in a successful marriage. When one partner despairs, the other shares hope; socioeconomic tasks are divided out; strengths are maximized, weaknesses protected. A successful union is a choreographed dance, rhythms in phase, which is far more efficient and economical and satisfying that the mere sum of individuals. There's a synergistic profit. That's true not only of a mating relationship, but also of society as a whole. Were it not the case, human society wouldn't cohere. And, in a correlated society it's easier for any one member to keep his own frame of reference flat. Flatness becomes the path of least resistance.

G: But in non-tribal society, and in unsuccessful marriages, the natural condition of correlation breaks down.

H: Because the individuals use cognitive symbols to make observations that alter their point-of-view and thus separate their universes. In marriage, the observations are usually literal—critical and often spiteful. But adultery must also be viewed as an act of observation, because it assigns the adulterer a separate identity, which obviously

241

fractures the marital correlation, and it's especially destructive because it precludes re-correlation. Short of adultery, re-correlation through intimacy is almost always an option.

G: How about the breaking of correlations in general society?

H: Greed and deceit are the destructive forces there. Any act of greed or deceit involves a separation of identities. These behaviors, you know, are fundamentals in capitalism, and for centuries were deliberately restrained—or at least contained—by religious and ethical considerations. Not any more. American society has effectively eradicated belonging and true community, and even intimacy has been nearly expunged. Greed is shameless, and deceit is so engrained we don't bother to notice it any more. Look at television. It is one hundred percent deliberate deceit in the service of greed. It doesn't even pretend to truthfulness or art. It is a twenty-four hour, five hundred channel, paean to greed and deceit, a stupendous monument to correlation-breaking.

G: But in any modern society, when you have non-egalitarianism, you're going to have prevalent correlation-breaking. That's in the nature of the political state, starting way back with...

H: Assyria, perhaps? There was a definite break with old ways in that empire, an emergence of ruthless individualized consciousness.

G: So, what can be done? Enforced egalitarianism doesn't work. Communism has proved that, I think.

H: No, you can't force flatness on a symbolic reality.

G: So one must ask the question, can we have our complex modern society and still have a single shared universe?

H: Only if each member of the society recognizes the need to keep his or her personal frame of reference flat. The question you're essentially asking is whether society can have a collective and correlated consciousness while all the members have individual consciousness. That would be a first in history.

G: And wouldn't such a society, or utopia, be static?

H: Perhaps. I would hope that society could hold a flat frame of reference and still have progress. But it's an academic question. We must return to flat frames of reference, on personal, socioeconomic, and political levels, if we are to survive, and you simply aren't going to have flat societal frames without a healthy degree of correlation. The two go hand in hand. Correlation is a natural process, because any interaction tends to correlate the participants. But correlation in our society is ruptured as fast as it sets in. We're going through a difficult stage in evolution here, with this full-scale emergence of individualized consciousness that's reaching its peak in modern society, and we still have no cognitive paradigm to substitute for the traditional constraints on behavior that were furnished by tribal life. The rupturing of correlation happens not just because people constantly make conscious observations, but also because they're using cognitive symbols to support distorted ego identities. People walk around full of tensions that artificially flatten their distortions. In America today nearly everyone has a distorted frame of reference. And how can you maintain correlation with someone who's chronically distorted? You can't, not without being distorted yourself. Yet, without continuous correlation, society can't have internal balances. It slowly entropizes, and desperately steals energy from wherever it can. The violence that the modern economic giants are doing to the rest of the world is horrifying. To weaker nations as well as to the ecosystem. And that societal corruption is mirrored in the relationships of individuals. If it goes on, we will have totalitarianism or an apocalypse. So, we really don't have any choice. The question of stasis really is an idle one. If it came down to that, wouldn't you rather live in a healthy, integrated society that doesn't progress than see humankind go extinct?

----Twelve----

Grana: Throughout our conversations you've alluded again and again to a single set of symbolic transformations. This set governs all change in reality—any type of reality—and is derived from the basic blueprint for creating Somethingness from Nothingness. Are we ready now to talk about it?

Heidrich: The set of transformations has two aspects. The first is: observation converts potential energy in the observed form into entropized energy. The second is: synthesis converts that entropized energy from the observed system into the potential energy of a new system.

G: Which would be a cyclic process. And the direction from which you perceive the transformations occurring depends on your point-of-view.

H: Correct on both points. From the point-of-view of the old system, observation is slowly eroding it into wayward sparks of kinetic energy. But from the point-of-view of the new system, the loose bits of kinetic energy are synthesizing into its coherent form.

G: And the observations which are eroding the old system, those observations are being made by the old system.

H: More or less. The system, by observing, is analyzing itself and its world into bits of information. The analysis provides information about the system, and is integrated into new symbols, but it also generates entropy.

G: Now, the old system is a symbolic reality, and that is equivalent

244

with an interference pattern. Since the interference pattern is composed of balanced dimensional warps, the entropy generated must be of warp values not held in balance.

H: When you make an observation, you collapse a qwf—a quantum wave function. You collapse a qwf by reducing the number of spatial dimensions that express that qwf. The spatial dimensions you have eliminated instantly become free values. They are no longer held in balance. They become kinetic and entropic.

G: But not for long.

H: No. In fact, we can safely presume that the entropized warp values immediately belong to a new frame of reference, or perhaps to many potential frames of reference. But from the point-of-view of the observer who loosed them, the values are out there somewhere running free. Unmanageable. When you're making observations, you're playing around with the balances that compose your frame of reference, altering them in ways that you can't even guess. The universe itself hasn't lost any balances, of course, but it may be a good while before you discover what the new interference pattern is.

G: And if you don't make any observations, then you won't know anything about the world you're in. We can't see the sun without observing photons from it, and we can't understand the atom without smashing it. We can't even ask another person a question about himself and hear him answer it without subtly altering who he is. What I'm getting at is that we're damned if we do and damned if we don't. We have to make observations to even be a point-of-view, but whenever we make observations we're changing whatever it is we're observing.

H: True. As you say, a point-of-view cannot even be, unless it makes observations. As human beings, we must absorb photons, we must breathe oxygen, we must digest food, or we won't exist. But, in an even deeper sense, our fundamental composition is of observations. We are the product of twenty billion years of observation gradually becoming symbolic form capable of self-recognition. In that sense, even the balances that give substance to our reality are nothing more than observations piled on top of earlier observations. It follows, then, that the history of the universe since the big bang is no more than the development of point-of-view, in progressively more detailed observations.

G: A history punctuated by syntheses that divide each old frame of reference from each new one.

H: And, I suppose that if you could talk about what's transpiring with heightened objectivity, you wouldn't even take note of old frames and new frames. You'd perceive only an ongoing flow of change. Only from the inside, from a subjective point-of-view, do there seem to be abrupt discontinuities from one frame to the next. This set of trans-formations, it's a set of two aspects because any given point-of-view is a static gestalt. You can't describe change without having two points-of-view—a prior one, and a subsequent one. The prior view describes change in one way. The subsequent describes the same change in quite another. So, instead of a single type of change, you have a set of transformations, with one aspect for the entropizing frame, and one for its successor. The analyzed and the synthesized. The relationship between the two is expressed in energetic exchanges: potential to kinetic, kinetic to potential.

G: Thing, change, thing.

H: Succinctly put. Thing being symbolic form, change being energy conversions. And these are the fundamental processes of reality, any reality. Any reality is a symbolic model of an original Nothingness, and must remain flat against time if it is to be an accurate representa-tion. So, change always occurs in a way that maintains the flatness.

G: Physical reality never loses its flatness. Abstract models of it, however, become distorted until synthesis flattens them again.

H: Yes, that's true. But again, if you could see it all with heightened objectivity, flatness is never really lost. When you're dealing with two points-of-view separated only by time, the distortion is relative to them. It's the act of dividing steady change into two time-separated points-of-view that gives an appearance of distortion to the prior frame of reference. The distortions exist in relation to a particular point-of-view. Which doesn't make them any less significant, because we can't live in total objectivity. We're subjective by our very design.

G: But what do we mean by *we*? What criteria define what is a point-of-view?

H: Well, we tend to suppose that point-of-view implies some sort of cognition, don't we? But in unireality, any sort of symbol, or symbolic milieu, constitutes a point-of-view. After all, a symbol is a representation, and a representation is a point-of-view. Since the universe is entirely symbolic process, it is axiomatic that it is composed entirely of points-of-view, and nothing else. Points-of-view that are in constant change, true, points-of-view that correlate and overlap and evolve, but still nothing but points-of-view. Every atom has a point-of-view, every molecule, even every star system and galaxy. Granted, such points-of-view are not self-aware. They are less sophisticated than a living creature's, less sensitive to environmental changes. And, since we have dispensed with the notion of an objectively real universe, we must presume that the differentiation degree of the point-of-view corresponds to the differentiation degree of the universe perceived. That is, the universe of an atom is consistent with the universe that is yours or mine, but not identical with it. To the atom, most of history is entirely irrelevant, and so for the atom it simply hasn't happened. To the atom, nearly all present events are irrelevant, and so they don't occur in its universe.

G: And all these points-of-view of the universe are overlaid.

H: And correlated, at least until an observation is made that requires them to separate.

G: And any one of these points-of-view, above the level of the proton and electron, will perceive its frame of reference entropizing and eventually collapsing. Because, no matter how well designed the symbolic form, its existence will become cyclic. It will follow a developmental course toward collapse, no matter how hard it struggles against it.

H: The general rule is that the longer synthesis is postponed, the more radical it will be when it happens. The longer that synthesis is avoided, the more entropy is generated and the more distorted the underlying cycles are. The new symbolic forms emerging from the wreckage are going to be dramatically different.

G: May I then summarize? The theory of unireality is this: reality is symbolic; the blueprint for structuring a symbol also tells how something can emerge from Nothingness; the way it emerges is by a single set of transformations; and, such a reality exists only subjectively, for a point-of-view within it.

H: To that I would only add that reality at any given time must be perpendicularly flat against the flow of time. Symbolic change is governed above all by the requirement that flatness is maintained, or that it is restored. It isn't possible to overestimate the significance of this fact. From it derives a standard of reference that cuts across all human behavior, all cultural mores, all possible worlds. Without it there would be no coherency to reality; it wouldn't make sense physically or cognitively. Whatever morals or laws or ethics humankind devises acquire their meaning from this tendency toward flatness.

G: All of what we've said in these conversations is based on one supposition—that the universe is self-enclosed. In other words, that it doesn't exist objectively, and that it amounts to nothing. I grant that you marshal a couple of good arguments that this is probably true, and the idea does have an esthetic elegance, but what if it is wrong? Where does that leave the theory of unireality?

H: First of all, the supposition has no worthwhile alternatives. If you want to suppose that the universe is not a symbolic model of Nothingness, the burden falls on you to suggest what else it might be. And while you might think that sounds easy, it isn't. Any other paradigm is incomplete. If the universe doesn't cancel down to Nothingness, then it is a Somethingness. Conventional cosmology claims it to be the remnants of an explosion of a point with infinite density of energy. Religion claims that a pre-existing god created it in an act of will. In either case, you've sidestepped the question of what it is, by positing the existence of something from which it came. From where did that point of energy arise? Where did the god come from? And even if you dismiss that causality question, the paradigms that develop from supposition of a prior entity are riddled with inconsistencies, paradoxes, and lapses of logic.

G: Okay.

H: Secondly, when we work up a paradigm we are making a model. True? An abstract framework that supposedly represents reality. All right, if the universe is not a self-enclosed zero-sum, then you cannot describe reality with a single paradigm. It's logically impossible. If there is even one open hole in the universe—be it time, energy, or god—then there can't be a structural dynamics that operates on all phenomena. Physical reality, transactional reality, psychological reality—these would not share the same dynamics. Without a single paradigm, there can be no theory of reality at all. There might be a theory of physical reality, a further theory of psychological reality, and so on, but there would be no way of unifying them in a single theory. Given a phenomenon with aspects in physical, transactional, and psychological reality, it could have three completely different meanings according to three completely different sets of dynamics. Meanings which might even be contradictory, and which certainly wouldn't be synonymous. For example, why should a concept like *honesty* have not only the same meaning in all three aspects of reality, but also the same implications? Why should honesty lead to positive value in three milieus of reality that operate by three different sets of dynamics? A universe that can't be expressed with a single paradigm shouldn't behave like that. Only a universe with a single ubiquitous paradigm should behave like that. And, if there's only one paradigm, then the universe must be emerging from Nothingness, and self-enclosed. If it is not, then we may as well throw up our hands in despair, because meanings will be utterly mercurial, and we're never going to find a solid place to stand.

G: I follow that.

H: Thirdly, the question is not whether the zero-sum supposition is correct or incorrect, the question is does it lead to a better description of reality than we now have. I submit that it does. It reveals pattern to what otherwise is unorganizeable information. It offers a set of transformations by which to make general predictions of future events. And, it integrates all old paradigms into one new one. For example, physics now has two very powerful theories: relativity and quantum mechanics. But the two are mutually incompatible, as they're now understood. The only way to unify them is to find a new theory that can contextualize them. The theory of unireality does so. It has the same effect in other disciplines of knowledge. Not saying people are wrong about this, wrong about that, but rather defining a structure within which what they believed to be true is still true. The theory synthesizes what is already known, and this, I should think, is perhaps the strongest argument in its favor. It does not contradict, but instead supports, previous knowledge.

G: By building an overarching structure that contains other knowledge.

H: Yes.

G: Knowledge that until now has been fragmented.

H: Yes. What are the things that people argue about? Is there a god? Is Freudian theory valid? Should women have the legal right to abort a fetus? Are we in a recession or not? Are corporate takeovers healthy for the economy? Who should pay for childcare? Should we allow compulsory drug testing? Does pornography lead to sexual violence? Where are we to put the homeless and who foots the bill? All such issues are issues because the positions taken are based on inadequate structure. At the heart of most if not all these issues are contradictory propositions, and they are insoluble as stated. They will remain insoluble until the contradictory propositions in them are contextualized by an overall description of reality. They are insoluble precisely because our current knowledge is fragmented, and no one is seeing the overall picture.

G: Perhaps the disputants will prefer that old-fashioned way of seeing things, even if the theory of unireality is correct.

H: Perhaps for now. But if unirealism is indeed a synthesis of what is known, then it offers an almost inconceivable energy of fusion, by binding into equilibrium the entropy losses that result from our distorted frames of reference.

G: And if unirealism is accepted and integrated, by the individual or the society, then the issues you've cited would resolve themselves?

H: Given a single paradigm in a self-enclosed universe, contradictory propositions are impossible. There's only one structure to what things are and how they change, so you can only use one set of terms. It isn't possible to define a problem in two non-translatable sets of terms—which is what goes on with these typical issues.

G: Why are the contemporary sets of terms non-translatable?

H: Okay, depending on which world-view you've settled into, you see things through some primary epistemological lens. Theologians, in the deep silence of their search for truth, analyze events and questions with terminology like "god's will" and "sacred/secular." Then, the women's rights groups evaluate events and questions with terms like "rape" and "dominance" and "equality." Psychiatrists use terms like "libido" and "transference." Therapists and educators have theirs: "inappropriate behaviors," "validation," and "vulnerability." Then in march the economists with their jargon, and politicians with theirs, and fundamentalists and neo-Nazis and militarists and activists and terrorists and so on, and so on. Each group has a framework of systemic dynamics that defines their terms. But no matter how well conceived the overall framework, it isn't anchored in anything ineluctable, and it isn't tied to the frameworks of any other group. Worse yet, it isn't even tied to physical reality, although it usually purports to be. If you examine the framework carefully, you discover that at its edges its connection to physical reality just sort of vanishes. Every single one of these world-views is a miasma of floating abstractions. If they aren't grounded, how can you translate them one to another? If you're asked to demonstrate a meaningful connection between *ego* and *trade barriers*, you can't do it. What does the term *rape* mean in a context of, say, epidemiology? Well, as long as knowledge is in a fragmented condition, and these structures are disconnected, nothing. And when two or three of these groups collide on an issue the debate is farcical. Can a man rape his own wife? The fundamentalists take one position, the feminists another, the judiciary a third, the buddies down at the neighborhood bar a fourth. And, how are they ever going to resolve the question when they're using conceptual structures that cannot be relationally diagrammed? Conceptual structures that have no common foundation?

G: But if the universe is zero-sum with a single blueprint governing how things change, then two groups cannot attack an issue with contradictory propositions?

H: Not unless they deny the validity of the premise. And if they do, the burden falls on them to either prove that there is no single paradigm or that the one offered here is not a valid one.

G: It seems to me that man's nature is inherently quarrelsome. We can't hope for an end to bickering, to disputes, to war.

H: Perhaps not. Still, when two football teams meet on the playing field they do agree on a single set of rules. The dispute remains, but they have a methodology for resolving it: whichever team has accumulated the most points by the end of playing time wins the dispute. But, if you send a baseball team onto the field against a football team, letting each side use its own equipment, how is the dispute ever going to be settled?

G: Which raises the question of how disagreements have been historically settled. By military and economic violence, I think.

H: When the issues are sufficiently polarized, yes. The abstract disputes are ultimately resolved in physical reality. That violence would be unnecessary if the arguments were grounded properly to begin with. I would suggest that, in the future, disagreements might be resolved by tracking abstract ideas through the transcendental structure of unirealism. We might need a world supreme court to evaluate the opposing contentions, and render a decision, but if there were general accord as to the validity of the single paradigm, couldn't we hope for some success with that approach?

G: And that is an optimistic note on which to end these conversations.

H: The theme we have emphasized over and over is that what we perceive does not objectively exist. To understand what perception is, therefore, is to understand all reality. Because, if there is nothing objective to perceive, then our technique of perceiving *is* the nature of reality.

* * * * *

A Few Last Remarks

With reluctance I have agreed to write a short commentary on how the world looks today through the eyes of the unirealist. The reluctance is not because I lack opinions or the wish to express them, but because the media today is already full of opinions about the state of the world. Too much of what is said is simple foolishness, and the rest is intelligent but largely irrelevant. I don't know that my opinions are any better, and even if they are I doubt they will carry any weight; opinions seldom do. To me, it is the theory that is important. Opinions can safely be built on solid theory, but when theory is built on opinions—as it so often is—disaster tends to follow. Opinions should take a back seat to theory. But, it has been pointed out to me that my theoretical ramblings are so dense as to be inaccessible, and so abstract as to be almost useless. This hurts my feelings a little, but I have to grudgingly admit the point is somewhat valid. So I will try to draw theory into the real world, and limit my exactitude.

* * *

Let me begin by observing that on that whole I believe human beings are extraordinary. It is not hard to imagine that we are some fantastic fluke of evolution. Even as animals, before we learned how to use symbols, we were probably pretty exciting to see. But this emergent phenomenon of consciousness! —It really has been a remarkable fifty thousand years. We have accomplished the most amazing feats, and if there are aliens out there watching us, I'll bet they find us fascinating.

The problem is that we don't know what the heck we are doing when we use symbols for thinking and communicating, and we are constantly being tripped up by our ignorance. Not society alone, but individuals as well. For example, the way we get confused and forget that the ego is merely a symbolic model or analogue of the true self. Or the way we get carried away with religious zeal, not realizing that religion is a by-product of our use of language. So, although we are creatures of astonishing intelligence, nobody gave us an instruction manual on how to use symbolic thinking, and we are blundering about as if blindfolded, and may destroy ourselves and our fellow creatures before we figure out how it works.

The fact is that technology is not going to save us. Nor is politics, nor is religion. We need to understand what it is that we are doing with symbols. We need to understand that they are just that—symbols—and that they will never perfectly describe reality and that we have to take that into account, always. Otherwise we end up burning people at stake because symbolic logic says we must. My wife and I have a running argument about this. She believes that if everyone would learn to listen to and trust the inner self, the world would be healed. I counter that that simply isn't going to happen, our symbols are too powerful and faith in them too absolute, so symbolic logic has to be de-legitimized first. The argument is never settled. I go on working to explain the ins and outs of symbolic reality, and she goes on her way shining the light of her inner self on the world. We all do what we can do, try to be what we are. I wish her way would work, but I doubt it will by itself. I think we also need a new paradigm to explain reality.

* * *

A few years ago Jared Diamond came out with a book entitled *Guns, Germs, and Steel*. The idea he advanced was that Europe came to dominate the planet because of those three factors named in the title. I read some of it, enough to recognize that he argued a convincing case. But I think that language played a significant role in Europe's ascension to power that Diamond overlooks. The languages that derived from Proto-Indo-European (the hypothesized granddaddy of all European and West Asian languages, which seems to have arisen in Asia Minor or the Caucasus region) all emphasized causality, because of their syntax. The effect of this subject/verb type of syntax may not have been especially pronounced when PIE was only spoken; its highly inflected nature may have mitigated the inference of causality. But as PIE spawned its host of daughter languages, the degree of inflection in most of them faded. Word order became more important, which put more emphasis on the syntax. When writing was invented, it had a further dramatic effect on linguistic cognition, because the left hemisphere was far better at reading than the right one was. The more that writing was used, the more dominance the left gained. Because the left is especially good at sequencing, the bias inherent in the subject/verb syntax was amplified. Causality, which had until then stayed in the cultural-cognitive background, asserted its supremacy in symbolic descriptions of reality.

This was especially true in cultures with phonetic writing, which had by then spread throughout the Mediterranean region. Other languages developed written forms that were hieroglyphic and pictographic, and reading these required active participation by the right hemisphere. The left did not take control. That is why even today the Chinese and Japanese cultures emphasize causality far less than the western world does. English itself is virtually stripped down to the paradigm of causality, more than any other extant language. It has lost most of its inflection, and depends on a subject/verb/object word order, driving home the point that what matters more than anything else is cause and effect.

The hidden assumption in Jared Diamond's view is that human progress is a given, and that some cultures just advanced faster than others because of their environmental advantages. But progress is not a given. Healthy cultures resist it, because progress is like cancer, relentless once it begins to spread, and it feeds on the lifeblood of natural systems. Progress is what happens when stasis fails. It is a consequence of causality unleashed.

When reality is perceived primarily in terms of cause and effect, cognition is taken over by logic, which undercuts the cultural resistance of traditions and rituals and ceremonies. Causality reigns supreme, and more holistic considerations are robbed of legitimacy. In emphasizing causality, we carelessly turn the magic of our world into commercialized banality. We invalidate the virtues of courage, kindness, charity, dignity, and integrity. Ultimately, we lose even our sense of meaning.

That is not all. The syntactical bias toward causality can drive a society on a runaway course of development that turns it into a political state and ultimately collapses it. We in the West like to think that progress is a natural dynamic in human society. On the contrary, the norm for most of human history has been stasis, with remarkably little change over centuries or millennia. Usually, when a society finds itself a comfortable environmental niche, it seeks to prevent change. That is why tribal and rank societies have disaccumulation and redistribution traditions and rituals. Their aim is to keep any one member or family from getting too wealthy. Because, it is the perception of unjustified inequality that leads to trouble in a society, and the greater the tendency to think in causal terms, the greater the trouble the inequality brings. The syntactical bias is like a genie kept in an open bottle. As long it isn't possible to accumulate wealth—because of ceremonial rituals that take it away as fast as it is acquired—the genie's power is limited, and causality is impotent. But if the traditional ways lose force, or the syntactical bias is amplified by writing and literacy, the genie begins its insidious meddling. The static society launches into a runaway development. It turns into a political state, which is what we all live in today. The problem with political states is that they are inherently unstable because of structural inequalities. They allow accumulation of wealth and power, and their only hope of cohering against the unleashed genie of causal thinking is to somehow legitimize the inequalities. Some political states, like China and Japan, have been successful in that regard, but historically the track record is not promising. Always just under the surface is the potential for revolution or collapse.

Seen in light of this analysis, political states today face serious problems. Although the European nations have managed to achieve approximate stasis through socialistic redistribution of wealth, they face a growing danger that their immigrant populations, who *do* perceive inequality in opportunity and status, will break into open revolt. China, meanwhile, has embarked on a capitalistic experiment that is almost certain to delegitimize their traditionally unequal society. India, with its own class-stratified society, is also tempting the genie. Furthermore, both of these nations now use English as a virtual second language, and its syntax may loosen their historical cultural rigidity more than anyone supposes. In the US, inequality of wealth and income has reached staggering proportions and the cloak of legitimacy over that inequality is thin and threadbare. The implicit promise that every generation will live better than the previous one has been broken, although the government is desperately concealing that by borrowing billions of dollars daily from countries like China and India. When the public finally must settle up these huge accumulated debts, and the magnitude of the betrayal hits home, rich Americans may have to flee the country just as the nobles tried to flee France and Russia during their revolutions.

* * *

The question to be asked at this point is why has the inequality of wealth reached such grotesque proportions in the United States? As we can see from the above analysis, inequalities will appear only if the traditions that resist them somehow fail. Every healthy society has traditions that are in place for exactly that purpose. They may show up as rituals, as laws, as progressive taxation, or as formalized social obligations. Whatever the nature of those traditions, they act to restrain individuals from acquiring inordinate shares of money, power, or access to the economic means of production. They are necessary because individuals are biological systems, which are always designed to maximize their own energy inputs. That characteristic can be countered by altruistic considerations, and tempered by reciprocal sharing, but it remains a ruling dynamic in biological systems. When individuals band into social groups, and available resources are held in common, increasing one's individual input means increasing one's share. I will not say it is human nature to want more than one's fair share, because it is a systemic characteristic, not just a human one, but it comes down to the same problem—which is why healthy societies have traditions that keep the tendency in check. When those traditional methods fail, inequality quickly manifests.

Various kinds of events can cause failure of the restraints. Invasion by another culture, or even a relatively benign contact like trade, can disrupt the social order. Natural disaster or rapid shift in climate may occur. A critical resource may become depleted, or a new one discovered. Existing technology may be improved, or a new one in-

vented. Most often, the cause involves a significant increase or decrease in the overall energy input of the society—to which the social
order does not adjust as nimbly as is necessary.

So what was the change to which our country did not adjust, and
which caused breakdown of the traditional restraints on wealth accumulation? It isn't the first time this has happened in the US. In the
eighteenth century, cotton and tobacco trade enabled southern planta-
tion owners to become very wealthy while half of the population
dwelt in impoverished servitude. During the nineteenth century, new
industrial technologies did the same for railroad magnates, factory
owners, bankers, and monopolists, while reducing independent farmers and unorganized laborers to virtual slavery. Considering these
precedents, it is hard not to notice a parallel with the computer revolution. The wealth hasn't ended up all in the hands of the inventors of
hardware and software, although some have amassed tremendous fortunes. More so, the newly available wealth has concentrated in the
hands of those who recognized how to capitalize on the new technology. This also parallels the precedents cited above. In both previous
periods, inventors of the new means of production may have gotten
rich, but the ones who became richest of all were those who found
ways to siphon off profits from the new means into their own pockets.

At first glance this explanation for the gross inequality may seem
facile, if not wholly mistaken. There are prime alternative explanations—the financial and psychological losses of the Vietnam War, the
OPEC oil price hikes, the hegemony of the Republican Party—and
none of these should be discounted, because they played significant
roles. But even in combination these factors could not have transformed society enough to allow such massive accumulations of
wealth. It was the computer chip that made possible just-in-time in-
ventories, HMO's, media conglomerates, foreign call centers, over-
seas production by US companies, job outsourcing, the merger and
acquisition mania, junk bonds, arbitrage and hedge funds, derivatives,
and corporate raiders. All these were prefigured to some extent in the
pre-computer economic world, but they grew by magnitudes with the
awesome power of memory and calculation of the silicon chip. And,
they made fortunes for the few hundred individuals who took full advantage of their potential.

The computer chip injected a staggering amount of new energy into the US economic system. It is unfortunate that the new energy/wealth was not distributed more equitably, but maybe anytime the energy inputs of a biological system increase or decrease rapidly, effects positive or negative are spread unequally. Supposedly, the decade of the Great Depression created more millionaires than any other decade in US history, at least until recently, but obviously most people went in the opposite direction. Inequality increased, just as it did during periods when the economic system's input energies dramatically increased. Systemically speaking, new energy floods into the deepest and most accessible channels first, and gradually disperses throughout the system until it reaches equilibrium. The problem with a biological system like the US economy is that greed and sense of entitlement interfere with any equalizing distribution, delaying it. It is an open question how long the present gross inequalities can be sustained by the powers that be, however. When syntax has a causal bias, the perception of inequality is a socially destabilizing force.

* * *

Paradoxically, although the computer revolution has generated vast new wealth, the country that pioneered it—the US—has gone bankrupt in all but name. At the beginning of the computer era, the US was the greatest creditor nation in the world, and now at the peak of that era the US is the greatest debtor nation. This would seem impossible. Productivity gains from the billions of chips installed through the infrastructure—not to mention productivity gains in other fields not directly caused by computerization, like agriculture and materials science—should have had the opposite effect. Even if the new wealth concentrated in the hands of the few, even if the many actually fell backward economically, the nation as a whole should be immeasurably richer, financially and in capital infrastructure. That is, after all, what happened during the two periods mentioned earlier, the late colonial and the industrial revolution. Despite inequity of wealth distribution, the nation gained tremendously in overall wealth. But not during the computer revolution.

Anyone who has lived in the US in the past few years can offer one good reason why—computers are often used in ways that waste more time and energy than they save. For example, as much as thirty percent of every dollar spent on health care in the US is consumed by administration costs. The typical doctor's office now has as many employees at workstations as providing health care. The use of computers has not only not reduced costs, it has multiplied them. And unnecessarily, in light of the fact that the government-run system of Medicare is administered at the cost of only a few percent. Nor is the health care sector anomalous. In every aspect of daily life, complexity has mushroomed, making too many tasks *more* time-consuming, and computerization has driven that trend.

Why should that be true? Why should a laborsaving, timesaving device sometimes make things take more labor and more time? How is it possible that such a marvelous new technology as the computer— surely as important an invention as the railroad or automobile or airplane—made this country poorer instead of richer? The answer lies in the way the computer has been used. If we consider the problem in the terms of system dynamics it becomes clearer what happened.

A system can generate energy—profits or gains—in two different ways, by analytic processes or by synthetic processes. The latter is preferable, because while both require an initial investment of energy, synthesis only has to be paid for once and is then basically free, whereas the overhead expense of analysis continues forever. That is, an analytic process can generate profits for a long time, but energy must be continuously spent to derive those profits. Synthesis, on the other hand, actually modifies a system so that overhead costs are reduced, which has the effect of freeing up a steady stream of energy at no additional expense. These are the only two ways of getting energy or profits of any kind. Energy processes in the real world may have elements of both, but usually can be easily classified in one category or the other.

We can think of the processes of analysis and synthesis in the simple terms of taking things apart to get energy, or putting things together to get energy. Analysis breaks bonds apart and releases their energy. Synthesis combines components and releases the leftover energy. For an example of the former, consider burning wood for heat energy. For an example of the latter, consider the fusion of hydrogen into helium, which gives off sunlight.

The computer chip can generate profits analytically or synthetically. Implanting chips in traffic signals has a synthetic effect. Installed, the chip maximizes traffic efficiency at the intersection, saving time and fuel, but requiring no additional expense. Robotics also works synthetically. So do sensors, and the Internet. So do communication satellites. In fact, our whole infrastructural environment is permeated with chips generating energy profits by synthesizing the dimensional components of the existing system.

On the whole, these computer chips generate great wealth. The wealth doesn't show up as more cash in the bank, or as more leisure time, as it could have. It shows up instead as novel forms of technology: cellphones, home computers that possess global communication capacity and desktop databases, portable music libraries, and so on. Arguably, much of the new wealth is misspent and overly complicates our lives, but that isn't the point. The point is that these technologies, whether appreciated or not, are the product of true new wealth generated by new synthetic processes.

In the background, however, the computer is insidiously devouring wealth more quickly than it is creating it. It is being used to capture energy by analytic processes, and in doing so it is gradually destroying productivity in this country. The runaway bureaucracy in the administration of health care is just the tip of the iceberg. Virtually every aspect of life has gotten dizzyingly complicated. This fact hits home every time a license or permit is required, a test must be passed, a report must be filled out, a contract must be signed, a request must be submitted, an inspection must be made, a degree must be earned, an ID card must be shown, a violation must be corrected, a change must be time-stamped, an update must be filed, a procedure must be pre-approved, a sign-in sheet must be initialed, a standard must be met, a fee must be paid, or a meeting must be attended.

What do all those interruptions of our lives have in common? This: every one is an attempt by an organization to generate, capture, or preserve a profit, by using an analytic process. Every one of these efforts in the long run costs the society more profit than it generates, captures, or preserves. And, every one has been enabled or necessitated by the use of computers. Many of them develop out of some sort of governmental regulation, but the corporate world contributes its share as well.

While this red-tape bureaucracy chokes off productivity in the microeconomy, sweeping changes are transforming the macroeconomy. Globalization of the labor force has undercut the working middle class, Social Security and pension funds have been plundered, profitable companies have been raided and stripped of assets, and the independent news media have been integrated into the entertainment industry. Growth in the economic sectors that really make things—manufacturing, mining, and agriculture—has stalled or reversed, while growth in the government, financial, and service sectors has soared. All of these trends are indicative of the analytic processes dominating in a US economy that was until a few decades ago thriving on synthetic processes.

With opportunities for producing meaningful goods and services disappearing from the economic landscape, too many individuals and businesses have resorted to analytic processes for extracting profits. This alarming tendency is manifested in multi-level marketing ploys, in the rising tide of donation requests from charities and political action funds, in spam e-mail and telemarketing, and in the recycling of used goods through cyberspace garage-sale sites. Needless to say, these are at best just clever ways of passing dollars from one hand to another. They do not produce anything new, and they do not synthesize what already exists. Worse yet, the people whose energies they spend must be supported somehow by the dwindling productive enterprises.

There are hundreds of billions of computer chips embedded in our infrastructure, regulating basic processes in utilities, machines, appliances, vehicles, and innumerable other devices. At the same time, computer programs have brought great laborsaving efficiencies to the fields of engineering and architecture, pharmacology and genetics, manufacturing and materials science, transportation and communication. The sheer magnitude of time and energy savings from all these syntheses is staggering. Where have all those profits gone? The answer is that whenever one person has been relieved of a laborious but productive job by a computer chip, some other person has been given an easier but non-productive job. Every time a factory has laid off one worker, the service sector has hired another worker. Every time a slot in an engineering school has closed up, a slot in business school has opened up. Every time a miner or farmer or craftsman or repairman is idled because a useful job is eliminated, someone else is put to work in law or government or Wall Street. That is where the profits have gone. Over the course of the past four decades, the US workforce has been gradually moved from jobs that produce profits by synthetic processes to jobs that produce profits by analytic processes.

I hope I have been clear enough about the difference between the two processes. A carpenter will frame a house and get paid for the work. An advertising agent puts together an ad campaign, and also gets paid for the work. Yet the former is a synthetic process of making a profit, and the latter an analytic one. It is utterly irrelevant that the business for which the advertising is designed sees a jump in sales revenue because of it. It is similarly beside the point that requiring sales clerks to smile and bid hello to all customers increases daily customer counts. It is irrelevant that a good legal department reduces the number of financial judgments against a company. To the business, to the store, to the company, these savings/profits are important, but not to the system as a whole. They produce profits analytically, not synthetically, and for the economic system as a whole they register as net losses, not net gains. Synthetic processes generate energy savings/profits because they consolidate, overlap, shorten, and combine dimensional components in the system. Analytic processes generate energy savings/profits because they differentiate, separate, lengthen, and break dimensional components in the system. Although any instance of energy production may contain elements of both processes, and although some degree of analytic process is unavoidable, systems that want to survive need to focus on syntheses as much as possible, and avoid analytic means as much as possible.

* * *

Not so surprisingly, the preference for synthetic over analytic processes within biological systems is a natural one. When all other considerations are reasonably equal, human beings are like other social animals, gravitating toward efficiency through cooperation. Society by definition functions synthetically; that is where it derives its evolutionary advantage. Reciprocating, cooperating, and sharing are all strongly synthetic activities. One need only consider anthills, wolfpacks, and beehives: their success is based on pooled efforts, with the resultant whole greater than the sum of the parts.

In a healthy society, members do not draw energy from other members. Rather, they draw it from that which is produced—which is produced cooperatively, to at least some extent. The distinction there is crucial, but apparently lost in contemporary economics, where a job is a job, and anything that is done during the performance of a job—no matter how nonsensical or pointless—is deemed to be productive. If a person is paid to dig a hole and fill it back in every day, that person is being productive—according to mainstream economists—and if he manages to dig and fill two holes in a day, his productivity doubles. Similarly, if a person invents a way to earn a living by legally stealing from others, that is also considered productive, so that if that person steals twice as much he or she is twice as productive. Anyone with any sense at all can discern that jobs like these would be at best non-productive and at worst counterproductive. But a large proportion of the US economy is composed of jobs that approach such absurd extremes—which indicates how unhealthy it is. In a healthy society, savings/profits are generated by synthetic activities, not by stealing, tricking, or compelling other citizens to hand them over.

What went wrong with the US? It gained its ascendancy in the world by synthesizing, not analyzing. Even its crime bosses, political machines, and robber barons were synthesizers, to a large extent. Why has the US turned its productive economy into a sea of analytic processes?

It is true that the states began as colonies whose natural resources were plundered analytically by the Old World, but immigrants to the colonies had much more in mind. They synthesized the resources they found, working them into farms and towns, mills and mines, foundries and factories. It was these synthetic processes of agriculture, mining, and manufacturing that made the US the richest country on earth. Nor was the accumulating wealth squandered in too many bloody wars, as had happened in Europe and Asia through much of recorded history.

The construction of this new economic powerhouse continued through the end of World War II. For understandable reasons, but with regrettable consequences, the US after the war never went back to its preferred isolationism. By then it owned most of the world's gold, and it went to work rebuilding war-torn economies—in enlightened self-interest more than altruism. In the process it became an empire with many far-flung outposts and ideological entanglements.

However, contrary to common wisdom, empires are funded by profits rather than providing them. By the end of the 1960's the US had already depleted its stupendous treasury, and was finding it difficult to simultaneously finance its military adventures and social programs. It had already, in 1963, stopped printing silver certificates. In 1965, the fateful decision was made to remove silver from minted coins. In 1971, the dollar was decoupled entirely from gold and since then has been backed by nothing except the strength of the US economy.

The long-term effect of these decisions has been disastrous. Never mind the consequences easily noticed—an inflation that has since devalued the dollar by nine-tenths, and interest rates unheard of previously. Far worse has been the drastic erosion of our system of values. Because, money is more than a medium of exchange, more than a store of wealth. It functions as the very foundation of our judgments and measurements and comparisons of value. Money projects an influence that extends far beyond commerce itself—it imposes order and meaning on all things. It establishes a base for determinations that frame even those many decisions apparently unaffected by monetary considerations; its reassuring stability undergirds ethics, purpose, hope, loyalty, kindness, charity, courage, and commitment.

This is especially true in the US, where money has been anointed with the revealing sobriquet of "the almighty dollar." The US was founded with no historical traditions, no particular racial identity, no national church or religion, any of which might anchor a system of values. The US started as an experiment in liberation, and was organized politically with the primary intent of making sure industry and commerce were efficient. If there was ever a nation requiring a reliable foundation for its currency, it has been this one, which may be one reason why the founding fathers included the provision in the Constitution that money could only be made of gold and silver.

When money is robbed of concrete meaning, the moorings on everything else are cut loose. The effect is subtle, but powerful because the subtlety goes unremarked. Little seems different at first. Merchants still accept dollar bills in exchange for goods and services, even though those bills no longer represent a quantity of gold, and vending machines accept all the new quarters and dimes, even though they are made of base metals. But deep in the psychological dynamics of the human brain, there is recognition that truthfulness has been replaced by fakery. Events that once took place in reality are now taking place on a stage. The words and the gestures are all the same, the characters unchanged, but in some vague way it all seems a sham. The worst part of this feeling is that it goes unspoken and unacknowledged by others. So it is denied. People give their heads a quick shake to dispel the vague sense that reality is somehow weaker than it was before, and then they go on about their business.

I know that anecdotes should not be used as proofs. However, an anecdote can illuminate a point, and here is one I would offer. Once we were visiting friends who lived in a very wealthy suburb of Los Angeles. We stopped at a convenience store, and while the others were inside shopping, I stood outside and stretched my legs. A boy of about nine or ten emerged from the store eating a snack. He got on his bike and rolled over to the curb. He leaned down and dropped a handful of change into the gutter. He started to pedal away, but I called to him and he stopped. I expressed curiosity at what he had done. Was it some local custom? Oh, the boy said, it just falls out of my pocket anyway, while I'm riding home. And as I stared at him, too stunned to speak, he rode off. Often I have replayed this scene in my mind, trying to find the exactly correct shade of meaning. What kind of neighborhood has no one poor enough to give change to? How rich do parents have to be before their children start throwing away any amount less than a dollar? Why didn't the boy buy himself a coin purse? Did the boy's parents ever notice he never had change on his dresser? And so on. But for illumination here, the question is this one: is it possible to imagine any period in US history when a child—no matter how fortunate of birth—would have thrown away a single silver coin? I do not think so. My impression is that that boy knew and accepted what adults still find difficult to admit—that our money is intrinsically worthless, and so in small amounts it is not worth bothering with.

Whether we admit it or not, however, we know it is true. And too many of us have tarred all that also once had value with the same brush. As money has no intrinsic worth, neither do integrity, honesty, courage, honor, and other virtues. Like money, they are props in the extravaganza stage production that is the US today. And, like money, they have no intrinsic meaning or value. They are no more real than tinfoil swords and cardboard shields. They are just as false as the stentorian declarations and conspiratorial whispers of actors in the footlights. Like the almighty dollar, they have value only to the extent that everyone agrees to pretend they have value.

It is not particularly controversial to say that this country has a crashed system of values. In fact, that topic is at the heart of more political discourse than any other is. The question is only why the once-vaunted system of values crashed, and for that there are primarily two answers. One faction blames the permissiveness and hedonism of the 1960's, which is supposedly still personified in the liberals of today. The other faction blames the corruption and moral vacuity of conservatives currently in power. But both of these are inadequate explanations. Even if the charge of permissiveness and hedonism is true, or even if the charge of corruption and moral turpitude is true, from whence did such tendencies spring? Changes result from decisions. When change occurs over a broad swath of individuals, it seems highly unlikely that those millions of individuals all made the same wrong decision. It is far more reasonable to suggest that an effective decision, made at the highest levels of the US government, warped the very underpinnings of US society and triggered a slow-motion crash of the system of values.

The choice to be made was between maintaining an empire and preserving the "lawful money" of the Constitution. It was decided to sacrifice the latter. The US kept and expanded its empire, and surrendered its system of values. But that is hardly surprising. It is a rare empire that knows how to bow out gracefully.

* * *

Religious fundamentalists may perk up at my frequent references to a system of values, thinking I have in mind a system of values from the Holy Bible or the Koran, or some other religious text. I don't. Religions come and go, but the system of values I speak of is inherent in reality. Religions at their best strive to express it, but typically make the mistake of trying to codify it as well. Such efforts are doomed to failure: reality does not specify acts that must be performed, or acts that are always prohibited. Neither does god—at least any god that exhibits credibility, by virtue of his or her consistency with reality. A society can make lists of required and forbidden behaviors, in the form of laws, but that is not the task of religion. In fact, the more a religion emphasizes rules, the less it is a true religion, because specificity works in opposition to spiritual appreciation. That is, true religion attends to the wholeness of things that is beyond time and space, and the more focus that is put on parts—which are expressed in terms of time and space—the less clearly can be seen the wholeness. One can study the brushstrokes of a great painting or one can step back and gaze at its beauty, but one cannot do both simultaneously.

The system of values inherent in reality arises from its dynamics—from the way reality works. If it is to be put into words, it must be stated in terms of how things happen, and not what things are. So it is inaccurate to say, for example, that adultery is wrong. What is accurate is to say adultery fractures marital intimacy in a way that is likely irreparable. Similarly, it is inaccurate to say that stealing is wrong. It is accurate to say that stealing is an analytic means of profiting, and that the entropy it releases into the environment will sooner or later find its way back to the profiteer, with negative consequences.

If a religion wants to make a list of disapproved behaviors, it should not claim the prohibitions are handed down arbitrarily from some deity. That entirely misses the point, diluting the force of the prohibition. The behaviors should be avoided because they are damaging, not only to society as a whole but also to the person who commits them—and not in the next world, but in this one. And, if a religion wants to make a list of recommended behaviors, it should present them as beneficial rather than requisite. Because although disapproved and recommended behaviors vary from one culture to another, they do spring from an unconscious understanding of the dy-

namics of reality. Attributing them to a deity is authoritarian thinking, typical of the primitive consciousness of the early political states.
The danger in such authoritarian justifications is that the demands become immune to reason and logic, which makes it all the more likely
that their utility will be perverted into cruelty and barbarism from
which there is no recourse. Torture, murder, and human sacrifice
then replace spirituality.

* * *

Differences between individuals, like differences between cultures, are often the result of how they use symbols. Early in these
pages I talked about how consciousness and thought were affected by
syntax and grammar, type of language medium, and type of writing.
The appearance of a culture—its structure, how it functions and
changes— is influenced as much by these factors as by environment,
history, and racial composition. Broadly speaking, all individuals in a
culture employ the same linguistic tools, and on a metacognitive level
approach and interact with reality similarly. The significance of this
fact is overlooked by ethnologists because it reminds them too much
of the ethnogenic theories that perverted science and politics in the
late nineteenth and early twentieth centuries, allowing or even en-
couraging the racial cleansing by the Nazis and the forced steriliza-
tion in the US. But we should certainly be able to entertain the idea
that language shapes a culture, and study the comparative effects of
different linguistic systems, without being tempted to assert the genetic superiority of one culture over another. Similarly, we should be
able to study how individuals use symbols differently within the
metacognitive context without being tempted to assert that one cognitive predisposition is inherently superior to any other.

I mean that sincerely. Although I am partial to my own style of
thinking, with its visual imagery, I appreciate the fact that other styles
of thinking are ingenious in their own way, and more useful than
mine, for other purposes. We are all specialists. We have all chosen—for whatever genetic or environmental reason—to emphasize
and depend on a particular style of symbolic thinking. In so doing,
we have also chosen to *not* develop other possible styles of thinking.

For each strength we possess there is a corresponding weakness. What we must therefore do is employ our strengths to maximum advantage, while limiting the downside of our weaknesses, and this dual approach can be better accomplished when the inverse relationship between strength and weakness is accepted. Carl Jung in his book *Psychological Types* explains the inverse relationship in terms of conjugate pairs—the thinking function repressing the feeling function, and the intuition function repressing the sensation function, and vice versa for both—and I recommend it to anyone who struggles with questions of why people are the way they are and act the way they do.

What Jung was not able to do was suggest a plausible physiological substrate for his four types of cognitive function: thinking, feeling, intuition and sensation. That I have tried to do, in my paper called "Consciousness and the Hemispheres." I won't go into it at length here, except to say that it appears to me that the four types of functions depend on the four areas of musculature used to hold the flickering tensions on which consciousness rides. However, the significance of the types does not end there, I don't think. It seems to me that they correspond to four different ways of representing our dimension-based reality.

* * *

As everybody knows, the reality we experience has four dimensions. Maybe there are others that are hidden from view, as physicists conjecture, but for our intents and purposes there are only the four: three of space and one of time. I don't want to get into the physics of reality now, because it would be too digressive. What is important to establish is that events not only take place in those four dimensions, but actually *are* entangled complexes of dimensions and nothing else. There are no *things* moving around *in* the dimensional matrix. There is just a very convoluted matrix, a sort of cat's cradle where the dimensions are string and reality is the pattern that the string makes. Matter and energy are just expressions of that dimensional matrix.

Any time we want to think or talk about what is going on in this dimensional matrix—this reality we experience—we need to find a way of getting hold of it. There is no way to manipulate it, no way to communicate it, as long as it is in its full four-dimensional splendor. So we cut it down to fewer dimensions. We dimensionally reduce it, which is to say we represent it. For example, say that a football coach wants his team to review their last game. He shows them a film of the game on a flat television screen—reducing the real game from three to two spatial dimensions. When a piece of music is recorded, it is represented in two or three spatial dimensions, depending on what device is employed. This process of symbolic representation is the essence of all communication, not to mention evolution and change, and I have written extensively about it in "Theory of Unireality" as well as "Understanding Reality."

When it comes to cognition, there is the same necessity for symbolically representing reality. It is not possible to hold onto a thought, to access memory, to consider or cogitate, without converting the flow of sensory experience into symbolic form—representing it so that it can be manipulated. However, there are four dimensions, and so there are various options on how to dimensionally reduce it. And here, I think, we discover the dynamics that underlie Jung's four types of cognitive function. There are different ways of dimensionally reducing reality into manipulable representations, and although everyone can use the four general ways of doing so, each individual emphasizes one of them, uses a second for support, develops a third inadequately, and represses a fourth.

* * *

It goes without saying that a lot of thought transpires at a non-verbal level, although to some extent it is harnessed with words. I am not referring to awareness—the flow of perceptual experience—but to consciousness itself—the skilled use of small tensions to direct and control neural patterns. It is in this kind of activity, this non-verbal processing of information, that the different functions come into play. The brain sends messages to somatic fields to set up tensions representatively, and proprioceptors return information about those tensions to the brain, which then modifies them according to how it wants to direct continuing thoughts. The brain is, in effect, using musculature as a sort of chalkboard; it makes notes, considers them, integrates its scribbbles, and erases them from time to time, to start again. This is not awareness, over which the brain has no self-control. It is what we call consciousness—the deliberate control of perceptual fields to facilitate thought.

As noted above, there are different options for dimensionally representing reality. Reality begins with four dimensions—three of space and one of time—and so it can be represented in three, two, one, or zero spatial dimensions. In each of these, time is still present; the representations are not static, but express time in their design. As I see it there are four options.

The first is the intuitive, which uses three spatial dimensions to make cyclic patterns. The second is feeling, which uses two spatial dimensions to make wave patterns. The third is sensation, and it uses one spatial dimension to make vibratory patterns. The fourth is thinking, and it uses zero dimensional points to make on/off patterns. Notice that in each of these the dimension of time is clearly present, so the representations vary only in the number of spatial dimensions to which the reality is reduced. These are the four elemental processes of pre-verbal thought. Of course, even when the brain is using only one of them at a time, the resulting neural patterns will be complexly layered expressions of the process. And, although the processes cannot be blended, the brain can switch rapidly back and forth between two or three of them to maximize its cognitive ability. At some point, the developing neural patterns reach a threshold level where the thought turns into an action, either vocalization or movement.

None of these four methods of pre-verbal thought is intrinsically superior to any other, although it is usually true that one is better suited for some specific purpose. If one is trying to understand a set of directions, thinking is most likely appropriate. For interior design or cooking, sensation is likely the wise choice. For writing realistic dialogue in fiction, feeling would be effective. For imagining a novel way to solve a problem, intuition may work best.

There is a trade-off between utility and completeness with these functions. The thinking function is the most useful, but it is also the least complete, having broken reality down into a collection of simple on/off bits. Sensation and feeling follow, becoming less useful but more complete. At the far end of the spectrum is intuition, with the most complex symbols, which makes it complete but imprecise and thus least useful in immediate reality. In other words, the fewer spatial dimensions a function uses to make its symbols, the more precise it can be and the easier to use in a situation. However, some information is lost when reality is dimensionally reduced, and the more reduced the representation, the more information is lost. The more spatial dimensions a function uses for its symbols, the more complete it can be, but while this has value in the long haul, it is difficult to manipulate usefully in the short run.

One point that ought to be noted here is that these functions exist as conjugate pairs. Feeling and thinking have an inverse relationship, and so do sensation and intuition. The more that one of the functions is emphasized, the more repressed is its inverse counterpart. Thus, any one can be used as strength, but it brings with it a weakness, because the counterpart repressed function is bound to surface in unconscious and troublesome ways. While one pearl of wisdom to learn from study of the Jungian personality types is that people employ different but equally worthy techniques for cognizing reality, it is just as important to understand that any preferred function has inescapable vulnerabilities.

* * *

In the pre-verbal thought, the brain is continuously inventing symbols, modifying and integrating them, erasing and forgetting them. These pre-verbal symbols are mediated by muscular tensions, which the brain employs as mnemonic aids so that it can control and direct the flow of awareness. But at some point—sooner for extraverts, later for introverts—the thoughts are expressed as acts: words are spoken or actions performed. Then the flaws in the symbolic representations affect the real world. It is then the trouble begins.

Achieving accuracy and precision with language is always problematic, because of the syntactical and grammatical—and terminological—biases that are certain in a symbolic system that attempts to accomplish so much. The bias problem is compounded when the user of the language brings to it personal cognitive prejudices, especially since users are almost without exception unaware that they have any such prejudices.

These kinds of problems fall into two categories. The first arises when a function is being applied in a situation where it is not the best option. Someone with a dominant thinking function may try to think a way through an emotional crisis. Or, a sensation type might try to use that function for explaining why another person is angry. The second kind of problem stems from the fact that the least developed functions are invariably repressed, and find devious ways to express themselves. Here we see an overlap of Freudian and Jungian theory; Jung's concept is very similar to Freud's idea of the repressed unconscious.

Often in the real world a problem from the first category combines with one from the second. A sensation type person may try to explain something logically, and in doing so allows a repressed intuition to assign a paranoid explanation to it. Or, an intuitive type challenged by an emotionally charged circumstance may strike out physically. A thinking type may turn to self-mutilation to cope with inexpressible feelings. A feeling type may respond to logical criticism with accusations about the motivation of the critic. And so on.

So, an individual trying to cope with reality seeks to take in necessary perceptual information, but since language is used to organize and store this information, to make it accessible in the future, it becomes contaminated with any linguistic biases, such as the implication of causality. For human beings the typical environment is social, so perceptual information tends to be abstractly symbolic and ambiguous, which makes the stored information less reliable. With preverbal processing of the information, it is further contaminated by the cognitive function prejudices. Considering that these three problem levels can interact detrimentally, it is a wonder that we get anything right.

* * *

At the start of this commentary I advanced the opinion that most of our problems originate in the ignorance of how we use symbols for thinking and communication. It might seem that there are more prosaic explanations for the many things that go wrong in society, but I find those lacking. It is commonly opined, for example, that human nature is responsible for greed, selfishness, hate, and pride. But we really know little about human nature; we only know about human nature after it is filtered through an array of symbols. Corporations are accused, money itself is accused, the dark lord himself is accused for the woes and tribulations of humankind. Childhood trauma gets its share of blame, religion is indicted, and genes selfish or aggressive are fingered as the causes of strife and suffering. These are surface phenomena, in my view, and attempts to ameliorate the human condition by addressing them is a matter of treating symptoms instead of the disease.

We naively believe we can describe reality, with its complex tapestry of four-dimensional flows, by using strings of simple abstract symbols. Not just experts, mind you, but everyone, regardless of level of education or proven competence. Let us suppose, however, that there are some people who have the most consummate skill with the twenty-six simple symbols of our alphabet, and that the words made of those twenty-six letters have fixed and precise meaning. Even in such an ideal set of circumstances, there is no right way, nor any complete way, to describe reality, because that which exists in four dimensions has to be represented in fewer than four. The cognitive functions can and will describe reality in the terms of frozen motions—cycles, waves, vibrations, and on/off pulses—and then assemble these symbols in some arrangement that approximates a flow. But no matter how skillfully executed, the arrangement will be a mere representation. And in reality, we are operating under far less than ideal conditions.

We might as well try to paint pictures of a carnival on sheets of cardboard, with fingers dipped in a child's paint set, and then fan the stack of cardboard sheets to create a description of the raucous bustle of a carnival in full swing. Even when this clumsy process brings results, we should never deceive ourselves into taking our descriptions too seriously. Nor should we fully trust our private thoughts, which can be crude interpretations and overly facile.

There are ways to cope with the conflict of having to use symbolic language without fully trusting its accuracy or conclusions. One solution that has proved helpful to many is meditation, the point of which is to place oneself in the flow of phenomenal awareness and not let symbols form, for an extended period of time; in that tensionless state one *becomes* reality, with no meanings assigned. Ritualized dancing and chanting and prayer are cultural first cousins of meditation, and serve the same general purpose, although they can be hijacked into the service of symbols. The therapeutic process practiced by professional psychologists is also an effective way to neutralize the power of inaccurate reality descriptions. Hallucinogenic drugs also can have a neutralizing effect. Finally, there is the simplest antidote of all, which is simply to always be kind. That is, or ought to be, the quintessence of any true religion.

However, those alternatives only help the individual who voluntarily seeks help. They do not put a stop to the outrageous and grotesque abuses of symbolic language that spill in torrents from our media, our institutions, our politicians, our corporations, and our government. Perhaps a fraction of it is the result of incompetence. Most of it, however, is uttered in a deliberate intent to deceive, to distort, to spin, and to obfuscate, for the cynical purpose of gaining profit, power, or advantage. There has never before in history been a culture in which truthfulness was so thoroughly stripped from all discourse, and if we were not acclimated to it, we would find it horrifying. Language has been the greatest achievement of humankind, and without it we are animals. It should be sacred, treated with respectful awe. The callous manner in which it is prostituted is an indictment of our way of life.

* * *

I don't want to be a world-improver. I want the world to be a better place, as do most people, but I do not think we can legislate truthfulness. The knee-jerk reaction of fixing problems by passing more laws and regulations is most often a regrettable one. When a government or a corporation tries to fix a problem, there are too often unintended consequences that make things worse.

The idea of mounting a full-scale legislative assault on deliberate deception is certainly tempting. However, compelling citizens to behave according to an agenda based on theory—whether that theory is religious like the Bible or the Koran, secular like Marxism or Maoism, or metaphysical like unirealism—never works. Of course, that does not stop self-styled intellectuals from trying. The neoconservatives who coaxed the US into Iraq were firmly convinced of the validity of their idealistic theory, despite knowing that history is littered with the wreckage of similar theories. It was the same kind of thinking that led to Prohibition, to the Cold War, to the Vietnam War, to the clumsy Great Society programs. World-improving fails so frequently because, in the first place, the theoretical reasoning is based on ill-considered assumptions, and in the second place because dictated rules tend to interfere with the efficiency of energy flows in a system.

If a theory is reasonably correct, it should not require an agenda. Political science theories are no different than physical science theories in this regard. No person was forced to accept the theories of Galileo and Copernicus, or of Newton and Einstein. Those theories were embraced because of their merits, unlike the previous Bible-based cosmology, which was enforced with dire penalties. So it should also be with theories explaining political society. If Marxist philosophy must be administered by force, it cannot be correct, and if tens of millions must perish before Maoism can be successful, it cannot be correct.

As far as the principles of unirealism are concerned, there is no need to encode them in law, even if there were a practical way to do that. But I do think they need to be understood and taught, and I believe that if that happens there will be a significant shift in how reality is approached. What I have done is provide a rational—and I think irrefutable—basis for truths and ideals that have until now been supported only by faith. Faith has proved a poor bulwark against abstract symbols; the abstract symbols not only create ego, tribal, and national identities, but also arm them with crude logic against which faith is ineffectual. Faith has never died—I don't think it can. But too often it has been backed into a corner by easy slogans: the only good commie is a dead commie, greed is good, manifest destiny, my country right or wrong, and uncountable others. Perverse logic has been used to justify genocide, torture, infanticide, massacre, slavery, and conquest. The fault lies not in abstract symbols themselves, but in our ignorance about what they are and how they work. I think once that curtain of ignorance is lifted, the power of perverse logic will diminish.

* * *

I would have thought that would be the end of my comments. I am informed otherwise—too much theoretical, too little practical, I am told. So I'll try something that I usually avoid because I find it tedious in other commentators. That is, I'll suggest a list of advisable changes:

1. Restore the basis of the value system. Give our currency the incorruptibility it had when gold and silver backed it. Fiat money invites a pernicious moral and ethical relativism. Even worse, it conceals the referential standard of reality, and allows nonsense and incompetence to reign.

2. Strip corporations of their personhood. While real individuals and societies instinctively seek synthetic means of producing energy, organizations and institutions do not. It is in their very nature to employ analytic means, which has a corrosive effect on the societal system. To assign corporations human autonomy and rights is to deliberately manufacture sociopaths and turn them loose on the world.

3. Allow debt to be taken on only if it will implement a synthesis that will generate sufficient energy savings to justify the loan. This applies to the individual borrowers as well as to governmental bodies. The costs of system maintenance—education, health care, and infrastructure repairs—should be drawn from current revenues. One side benefit of this rule is that it would stop virtually all warfare, since war is almost invariably financed by debt, and war is almost always for analytic purposes.

4. Stop financing the media by advertising. In fact, commercial advertising should be almost entirely banned. It should be permitted only in locations where a consumer would go looking for it—catalogs, stores, business websites, etc. Commercial advertising is analytic energy production run amuck. Using advertising to pay for our communication media cannot help but corrupt its content.

That should about do it. There are many things wrong with the world today, but once reality is understood for what it is, most other problems will go away by themselves. To defend their egos, for example, people inflict considerable damage. By legitimizing the true self and delegitimizing the ego—its symbolic representation— unirealism alleviates that cause of so much pain and strife. Or, consider religion. It has been a perpetual source of dissension that often culminates in bloodshed. But surely understanding that religious dogma arises from a syntactical overemphasis on causality will blunt the force of religious zealotry. The power of plausible explanation should never be underestimated. With better understanding most problems will fix themselves.

* * *

Although I believe that understanding reality is key to survival of our civilization—not to put too fine a point on it—I am not sanguine about immediate acceptance of the principles of unirealism. Everyday people won't object to them—not strenuously, anyway. The principles make too much sense, and validate much of what is already known on a gut level. The business world won't get up in arms against the theory advanced here, either. If it pays the ideas any attention at all, it will be only to figure out how to improve profits by focusing on synthetic means of improving bottom line numbers. Government? —Not likely there will be much opposition there, either, since nothing in what I have written will interfere with politicians getting re-elected. And, in spite of the casual way I have dismissed the whole premise of religion, I don't think there will be preaching from the pulpit against unirealism. That would be hard to do without actually teaching its principles, not a wise course of action when defending fantasy and nonsense.

No, the most furious denunciations of the ideas proposed herein will come from academia. There is too much here threatening to the little fiefdoms that tenured purveyors of the conventional wisdom have established for themselves. In particular, the quintessential theme of unirealism is that there is only one reality, no matter where one looks. The academic disciplines are not really separate—they are just aspects of reality under the overarching umbrella of unireality. I hope I am wrong, but I find it hard to imagine that professors who have finally climbed to chairmanships in chosen fields will appreciate being told that the studies to which they have devoted their lives are merely subsets of a larger field hitherto unnoticed by them or anyone else. I imagine, in fact, that what I have said in these pages will be ripped to the smallest possible shreds, and myself pilloried.

That of course will not make me wrong. No one with new scientific ideas has ever been greeted with genuine bonhomie. Academics are prideful, ambitious, egotistical, self-righteous, and vain, like the rest of us except maybe more so because they are paid less for doing what is so obviously more important. I believe that they are well intentioned. I believe they are liberal. I believe they are good people. I do not, however, think they are as open-minded as they picture themselves. I think that independent and original thinking was drummed out of them long before they got their doctorate degrees, and that they will hate my description of reality. I think they will pounce like jaguars on any hint of weakness in it—a claim that contradicts even scant empirical data, an effort to reconcile what several millennia of philosophy have failed to reconcile, a generalization that has apparent exceptions. Again, I do hope I am wrong. This is not about me. This is about the future of the human race. If we cannot understand reality we are doomed, because we are simply too smart for our own good.

So I offer a caveat, a qualification: I may be wrong about some things. In my heart, there is no doubt that the world is essentially as I have described it. That it is zero-sum, that it operates by rules inherent in its design, that it makes sense and the same kind of sense in every aspect of it. But there are many, many places in these collected works where I have sensed a truth and stepped off into empty space on faith alone. A small example is when I posited that the perception of inequality is what leads to change in a society. Is this true, always and without exception? I honestly don't know. I think it is very likely true, and on this basis I asserted it. Is it possible that I am somewhat and in some case wrong? Yes. Does that invalidate my general theory or my various other assertions? It does not.

Furthermore, these works have been written over twenty years, and there were nearly twenty years of research and synthesis before the earliest of them was completed. Hence, all of my ideas have crystallized over time. All have had prior incarnations. Although I now would phrase some earlier thoughts a little differently, they are sufficiently accurate to leave as they are. The careful reader may notice slight discrepancies, for example, between my statements in progressive works about human consciousness. I suppose this may be held against me, but accusations of inconsistency would be foolish. Metaphysical theory can never be perfect and never absolutely finished. Even the purest theories of physics are modified from time to time, and the subject I have tackled is far more complex than any of pure physics. The study of reality will never be completed by me or anyone else.

*   *   *   *   *

www.ingramcontent.com/pod-product-compliance
Lightning Source LLC
Chambersburg PA
CBHW031807190326
41518CB00006B/224